AF597648

PLAYFULLY RIGID

PLAYFULLY RIGID

SWISS ARCHITECTURE, GRAPHIC DESIGN, PRODUCT DESIGN 1950–2006

Edited and written by Claude Lichtenstein

Lars Müller Publishers

IN A MAGNETIC FIELD

This book is an attempt to divine a common meaning from many fragments. Since the fragments originate from different things, it would be wrong to try to assemble them to form a whole. Numerous individual works, each with distinct characteristics, backgrounds, and meanings, from three disciplines and from a period of more than half a century are brought into thematic proximity to one another. Despite this kaleidoscopic starting point, we nonetheless aim to present a whole – a book that not only contains individual tributes, but also explores the commonality of these works: the simultaneous presence of *Spielwitz* and *Klarheit* in an object, a work of architecture, or a visual representation – the manner in which they are clearly ingenious.* We are not concerned with a formal style, but rather another type of commonality that presumably rests on related understandings of design. What do we mean by this?

Clarity needs no explanation, even to an international audience, when attempting to elucidate a Swiss character trait. Sobriety, sense of order, control, mastery, correctitude, incorruptibility – these are, perhaps, several of the defining characteristics attributed to Switzerland (and, it could be elaborated: to all parts of the country and to all four linguistic regions – French-, Italian-, Romansh- and German-speaking Switzerland). Who shall say that these traits are but coincidence?

However, the concentration of such bourgeois virtues alone would result merely in a nation of green eyeshades and not the diverse cultural landscape that Switzerland embodies, now more than ever before. We may thus posit, in a scientific spirit, that there must be another force alongside the strong tendency toward order. What is this force, and where does it manifest itself?

The works presented and too-briefly honored here are more than mere evidence contradicting the cliché of Switzerland as a boring nation which prides itself on its correctitude. These works speak the language of a vital and, in the best sense of the word, original culture of design. They exude a self-confidence in the development of ideas that is free of any pharaonic or otherwise post-feudal pathos, but is nevertheless committed to standing up for their beliefs.

* Translators' note: The publisher and the editor chose "Playfully Rigid" as the title of the English version of this book. This introduction discusses the terms in the German title, "Spielwitz und Klarheit", Klarheit meaning clarity, and Spielwitz being an untranslatable coupling of "play" and "wit" which means "ingenuity while playing".

Almost one hundred works are presented here. Skeptics are bound to interject: How can one hundred examples stand for an entire country! Particularly examples covering a half-century and three professions; we could easily find one hundred counter-examples. We, in turn, could easily respond by doubling our selection, and so forth... But this debate would be missing the point– after all, it is well known that every vision of reality is a construction. A stickler for statistics would end up living in a virtuous Sparta of the Alps. However, we could not help but accuse this person of wandering blindly and dispassionately through the world. Individuals do not live a statistical ex-istence, but rather their own. And we do want to know what there is to experience and discover out there.

We now turn our attention to **Spielwitz**. We chose this term for its maneuverability off the beaten path of meaning. Occasionally, we encounter it in classical concert reviews; when the musicians are able to free themselves of the musical score and not simply play the music as written, but give it personal coloration and bring it to life, they have achieved a prerequisite for *Spielwitz*. *Spielwitz* is not playfulness. This is important. Playfulness in adults is usually a narcissistic pose, in which people's false sense of their own importance leads them to overestimate the audience's interest in them. *Spielwitz*, on the other hand, is subtler. It contains the important elements of perspicacity, of discovering that which is hidden, of controlling circumstances, and of adeptness at dealing with rules. There it is again, Helvetic virtue! Yes, but in this case it functions as an *antidote*, *Spielwitz* as a serum against pure correctitude. The French "esprit" comes close in meaning. It has to do with play and with the rules that designers give themselves. Whether they set the rules before beginning their work or – more likely – during the process is a secondary matter.

What is termed *Spielwitz* here is a method of giving meaning in a place and time of prosperity. It is an instrument many designers use to deal with a situation in which design reaches into many spheres of life, which most certainly is the case in a country as densely populated and permeated by the economic system as Switzerland. In a territory in which *everything already exists*, where there is something everywhere, a ruse is necessary for designers to create something of their own in such a thoroughly acculturated zone – to create something that does not simply replace existing matter with new matter, but rather fills it with ideal content not apparent at first glance. This is the ruse, and often a design motif seems to lie at the heart of it, that is to say parallel to the question of the primary function of an object. This is a subtle method of subversion in the name of perfection. The past twenty years have given designers a remarkable freedom to articulate themselves within an extremely controlled sphere of action as individuals who makes their own rules, for themselves and for their audience. Important examples of this can, however, already be found around 1950. If the discussion of design today is to be about more than a mere object and become part of cultural communication, then this game with rules of its own is an important element. Game-playing, yes, as a tool to combat perfection, and rules, yes, as justifications and thus also as a component of the skill which does not deny perfection, but wants to transform it. The *skill* of *Spielwitz* versus the perfection of controlled functioning – that is the inventive match apparently being played at this time.

And so, this, too, becomes clear: When we speak of *Spielwitz* and clarity, we do not consider them opposites. The two terms are not the endpoints of a linear scale on which a work stands closer to one end or the other. They interpenetrate one another. One can imagine an alloy in which the mixture of different portions results in a particular effect – creating something other than the individual properties of the initial elements (and thus also something more). If this metaphor is too materialistic, we suggest another one (our own): In our vision, *Spielwitz* and clarity stand for a polarity, and the field of tension which this polarity produces. This lends the objects a "vibrating energy" that is very important for their effect.

This hypothesis that *Spielwitz* and clarity have combined effects, over generations and despite the differences in the problems to be solved, is the common denominator of the compilation before you. We do not, however, feel compelled to present theoretical proof of its validity. This book is not a treatise. It proposes a new reading of the phenomena. We put it up for debate here, ourselves convinced of its relevance and also that this is a fundamental criterion for design quality by which the fashionable term "creativity" can effectively be put to the test. Aesthetics is not an exact science. Under the condition that the title "Spielwitz & Klarheit" is accepted as a proposal for a way to look at and examine things – and not as a robust doctrine – this book is meant to be understood as a source of inspiration.

In this collection you will find: two poles: *Spielwitz* and clarity; three métiers: architecture, graphic design, and product design; four linguistic cultures; Switzerland as a small but multi-faceted corner of world affairs and not as an island of bliss; a period of fifty-six years in which this investigation takes place: The question of structure arises here by necessity, when a continuum such as time is measured and evaluated. Like all other countries, Switzerland too has undergone major changes in the past fifty-some years; the starkly bipolar opposition of city and country has in this time yielded to the triad of city – agglomeration – country, within which increasing numbers of people move, ever farther and with increasing frequency. Manufacturing, in 1950 this country's greatest pride, has since largely been replaced by the service sector. This raises the question of how to organize the works presented, perhaps even according to an immanent model of development. How does one make time tangible? Which accentuations are plausible? This book does not wish to make any hurried attributions or fix things down, but rather set them in motion. Because of its heterogeneity, this material evades an approach that insists on searching for causalities. This is not a dragnet investigation for the paired terms *Spielwitz* and clarity. Rather we want to apply this criterion to inductive observation of selected examples – works chosen because they promise something in this regard. Free from the goal of a strict chronology, we nonetheless came to a periodization that we find convincing.

The five-and-a-half decades of our investigation can be divided into four periods that we could metaphorically call: *New beginnings after 1945 / Flying high in an economic boom / Turbulence in the jet stream of the postmodern / Soaring over lively landscapes.* Expressed in specific years, these are the periods 1950 to 1961, 1962 to 1973, 1974 to 1989, and 1990 to the present.

The dates make clear that these periods correspond with historical events: in 1961, the Berlin Wall was built and the Cold War heated up; in 1973, the oil crisis put an abrupt halt to the decades-long economic boom; in 1989, the Iron Curtain was drawn back and the Cold War came to an end; since then, we have observed and experienced deregulation, the fight for brand identities and personal profiles in a time in which goods, images and information travel around the world. These events are like cairns marking particular conceptions of the design and form-giving processes.

In accordance with these historical periods, the book is divided into four parts, entitled:

1. 1950 to 1961 Prototypes (début de série)
2. 1962 to 1973 Studies and concrete cases
3. 1974 to 1989 Reflections and alternatives
4. 1990 to 2006 Positions (fin de série)

Individual deviations from this chronology can be found in this book. Several examples are assigned to other periods than the year of their creation would indicate. Correct characterization counted more for us than the exact date.

We do not wish to go into detail on this point here, but there is no getting away from the thought that these periods also swing back and forth between the poles of orientation "form" and "structure". The first and third periods are clearly form-oriented, the second structure-oriented, while the current period seems to be investing great power of invention in the wish to work a structural interest into a marked interest in form.

Each of the four parts of the book is introduced with a short characterization of the period in question.

Is the connection between *Spielwitz* and clarity, the focus of this book, an exclusively Swiss affair? No. There are several other countries, societies and cultural groups in which it seems to hold equal significance. It could presumably be found in Italy, the Netherlands and Japan, though hardly in France, Great Britain, or the United States. We are, however, certain of the richness and variety in which this quality can be found in Switzerland, justifying an investigation of this kind. We have several suppositions as to why this is so:

1. General prosperity: Switzerland, as one of the richest countries in the world, does not have to struggle with making things work. It therefore can and must provide ample room for questions of design quality. A garbage can in a public space today is an object of design. Aestheticizing also holds a danger, which designers confront with various, sometimes subversive means.

2. In Switzerland, design is almost never the expression of effusive indulgence in form. Form that arises of its own accord and refers only to itself is seldom encountered here, and is even interpreted as a gesture of personal conceit, promptly meeting with disapproval. Designers generally avoid theatricality. Their discourse takes place on a level that often has a rational and comprehensible foundation. The "invention" involved in a particular work is often not evident at first glance; rather, the act of creation lies in solving a particular problem.

3. There is a tension between these inventions, in a narrow or broader sense of the word, and the Swiss people's highly developed need for security. These works always express a willingness to be nonconformist. During the process of design, this attitude also always entails a willingness to take risks. Frequently, this *creative risk* (the readiness to take risks while developing a solution) is a means to minimizing *risk while using* the object in question.

4. Too obviously expensive solutions are generally frowned upon by the Swiss public, which is astonishingly resistant to equating "design quality" or the “culture of building” with a high budget. In the present collection as well, the value of the works usually lies in their intelligent economy of means.

5. In architecture, as in building generally, the federal structure of Switzerland favors local initiatives. It allows productive working relationships between designers and local and regional authorities. In certain fortunate cases, the unconventional full exploitation of municipal autonomy leads to undertakings that would be impossible under a dominating central power.

We hope to spark discussion and further development of these deliberations, and that this book succeeds in drawing attention to the many creative impulses manifest in the shaping of our environment and society, and not only in Switzerland.

1950–1961

13 PROTOTYPES / DÉBUT DE SÉRIE

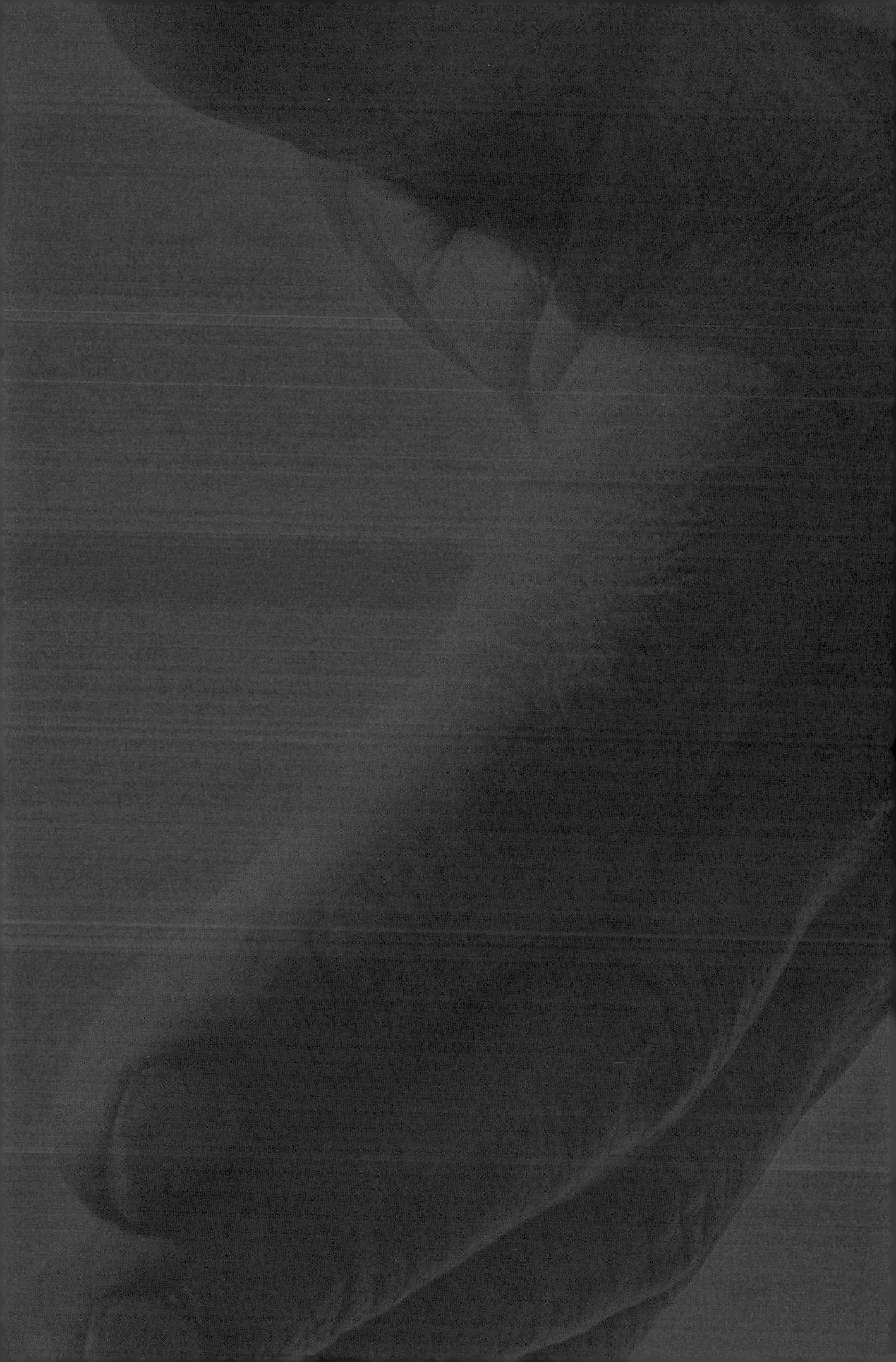

In retrospect, the first period resembles a deep breath and a sigh of relief. The war is over and the worst of the destruction in the neighboring countries has been repaired; even in Germany, Italy, Austria, and France, daily life goes on pretty well, in some cases astonishingly well. Switzerland entered the post-war race in 1945 with an enormous advantage. Although it slowly lost its special status in the years of normalization, it didn't notice – for one thing because there was good business to be done in the growing foreign markets. Thus the decline of its advantage didn't become general knowledge for quite a long time. The extent to which Switzerland was spared, not only in comparison to Germany, is *miraculous*. This belief in miracles is also the elixir of life nourishing the national myth of Switzerland as a "special case"; only the Swiss consider the miracle their just reward for superior industriousness. One could have realized (and some few did) that not only were considerable political opportunism and tactical skill on the part of the national government in play during the war, but also quite a bit of luck; but this insight did not come until much later, after the turn of the century, when it was laid open for all to see by the Independent Commission of Experts Switzerland – Second World War.

Around 1950, the post-war period ends and the *courant normal* begins. Those who are now young adults feel liberated after 20 long years: The period of global economic depression, totalitarianism (Hitler, Mussolini, Stalin), the pre-war shadows, the six long war years, and the war's end are finally over. What now? Life begins anew. One can finally think about traveling. Switzerland is no longer a peaceful island surrounded by war, but rather a refuge of intactness. The "Sekundenkelle", the large red second hand on the railway station clock is a beacon of this luck. (The Sekundenkelle was introduced by the Swiss National Railway around 1954; it announces not only the minutes until a train departs, but also the remaining seconds. In the course of time it will spread internationally; at the moment, however, this on-the-dot punctuality – and the sense of happiness associated with it – is still a Swiss specialty. It was conceived by Hans Hilfiker, an engineer who is honored in the coming pages by the inclusion of one of his works.)

Some of the works presented here remind one of throwing the windows open. In **architecture**, they can be found in impressive numbers, in **graphic design** as well. The case of **design** is not as straightforward. One reason for this is the simple fact that at this time no single word exists to denote that which will later be called "design". Experts speak of "industrial styling", the very progressive talk about "Gestaltung", the publishing world speaks of "applied arts", and the public has as yet no name for the phenomenon which has taken over their furniture, their kitchen and office appliances, and their interior decoration

– maybe at most they have found an expression for the formal idiom of their means of transportation, which they call “streamlined” when they speak of it with a racing pulse. But the term “design” is yet to be used intransitively (not the design of something, but design as a profession, a mission, a problem).

America is the distant victorious power which liberated Western Europe and ended the war. For most people, it is inaccessible: A plane ticket costs many months’ salary and who has the time needed to travel by boat? A week going, a week coming back, and then the stay in the enormous country – which one wants to see as much of as possible. But Europe wants to know what this America is all about. Some photographers are lucky enough to be commissioned to interpret and introduce the country in images (some of the big Swiss names: Werner Bischof, Emil Schulthess get good commissions from the extraordinary magazine *Du*, established 1941; the young Robert Frank, who emigrated from Zurich in 1947, is still unknown). Agencies are also producing a slew of images of America – machines, inventions, things like drive-in movie theaters and air conditioners, fast-food restaurants, spectacular fairgrounds, one record-breaking race after another in Salt Lake City, magazines like *Life*, and color movies. And of course jazz, America’s genuine contribution to music in the 20th century. This has to do with Switzerland inasmuch as many different fascinations are taking hold simultaneously and creating a modern attitude towards life. In architecture, Heimatstil is reaching the conclusion of its existence after a reign of twenty years. The days in which a building defiantly clung to the land, defending its here and now against progress, are coming to an end. The public is interested in new solutions with a claim to supraindividuality: parking garages, self-service stores, synchronized traffic lights on the main streets. In advertising, the pre-1950 genre image makes its exit, the charged freestanding object makes its debut. Portraying objects by themselves emphasizes not least the supraindividual within them. When one asks, “what’s new?”, the meaning of the question is: of all the new things which the modern world holds in store, which have finally found their way to us? That’s why we chose to title this era: Prototypes (début de série).

In **1954**, the graphic designer and artist **Carlo L. Vivarelli** designed the poster and the **catalogue cover** for the exhibition **"Nuove Forme in Italia"** at the **Industrial Art Museum** of **Zurich**. These catalogues were still octavos at the time, modest in length (in this case, 60 pages, 28 pages of text, and 32 pages of glossy paper for the reproductions). The cover was fashioned as a two-tone silkscreen print by the Industrial Art School in Zurich itself. ——— Vivarelli's cover design was inspired by the distinctive decorative approaches intrinsic to young Italian design (e.g. that of Gio Ponti). He utilized this approach by grading the green and light-red color fields and by superimposing them in two different places. Perhaps only then does it become apparent that these decorative patterns are, in fact, five replicas of the Italian tricolor, compressed into an approximate square. Its white center field is the solid blank cover paper. The catalogue's standard format width is divided into thirds, its page length yielding the tricolor. These are of course posterior explanations on the construction of this piece, not originally intended for the public. Yet for Vivarelli, they must have been devices in his genesis of clarity in
21 the presence and absence of tension.

In **1949**, the movie theater **"Studio 4"** received immediate international attention and recognition for its original interpretation of a cinema. Stage designer **Roman Clemens** from Dessau, working together with **Werner Frey**, the architect of the new office building, wanted to create not just a cinema, but a contemporary movie theater. Clemens found an admirable, creative and functional solution to the structural asymmetry of the theater auditorium, located across the entrance hall of Werner Frey's elegant office building – an irregular pentagon with three street elevations. Clemens made an offensive move by rotating the film room's geometry to the right and expanding it. He placed the projection room on the left side, giving the projectionist an emergency exit onto the courtyard. The left wall curves vertically into the ceiling, while on the right side, a frieze of black and white photographs (partly taken from *Life* magazine) meanders horizontally towards the back and ends, after curving twice, in front of the projection room.

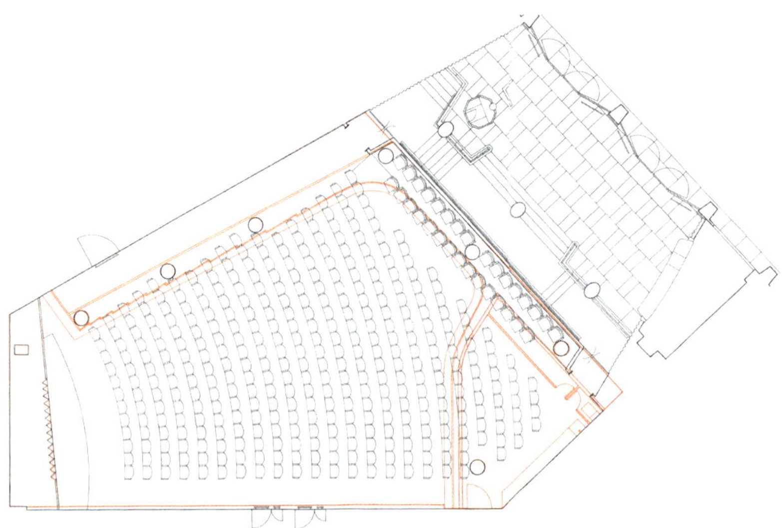

22/23 Movie theater Studio 4, view of the screen

Floor plan of the screening room and the lobby

Lobby with ticket booth

These two space-confining attributes, vertical and horizontal, are given an additional dynamic: the left wall and the ceiling through gray painted “clouds” and line bundles which purposely contradict the real vanishing point of the room by creating their own graphic perspective. To the right, the lowered entrance with its white vertical lines is dynamically enhanced through the smooth gradation of pictures above; and particularly through the “rotating” round pillars painted in a pattern of black and white blotches. Most exciting of all was the screen: a kinetic mechanism made of standing triangular prisms, which turned from black to white before the film began. Within seconds, the audience was led by kinetic means into the world of filmic plots. Even though Clemens’ patented kinetic screen was replaced by a static screen several years later – because the separation seams were disturbing – “Studio 4” remains an exciting place. Since 1983, it is an important institution: a communal film platform boasting an excellent program.

Screening room with photo frieze

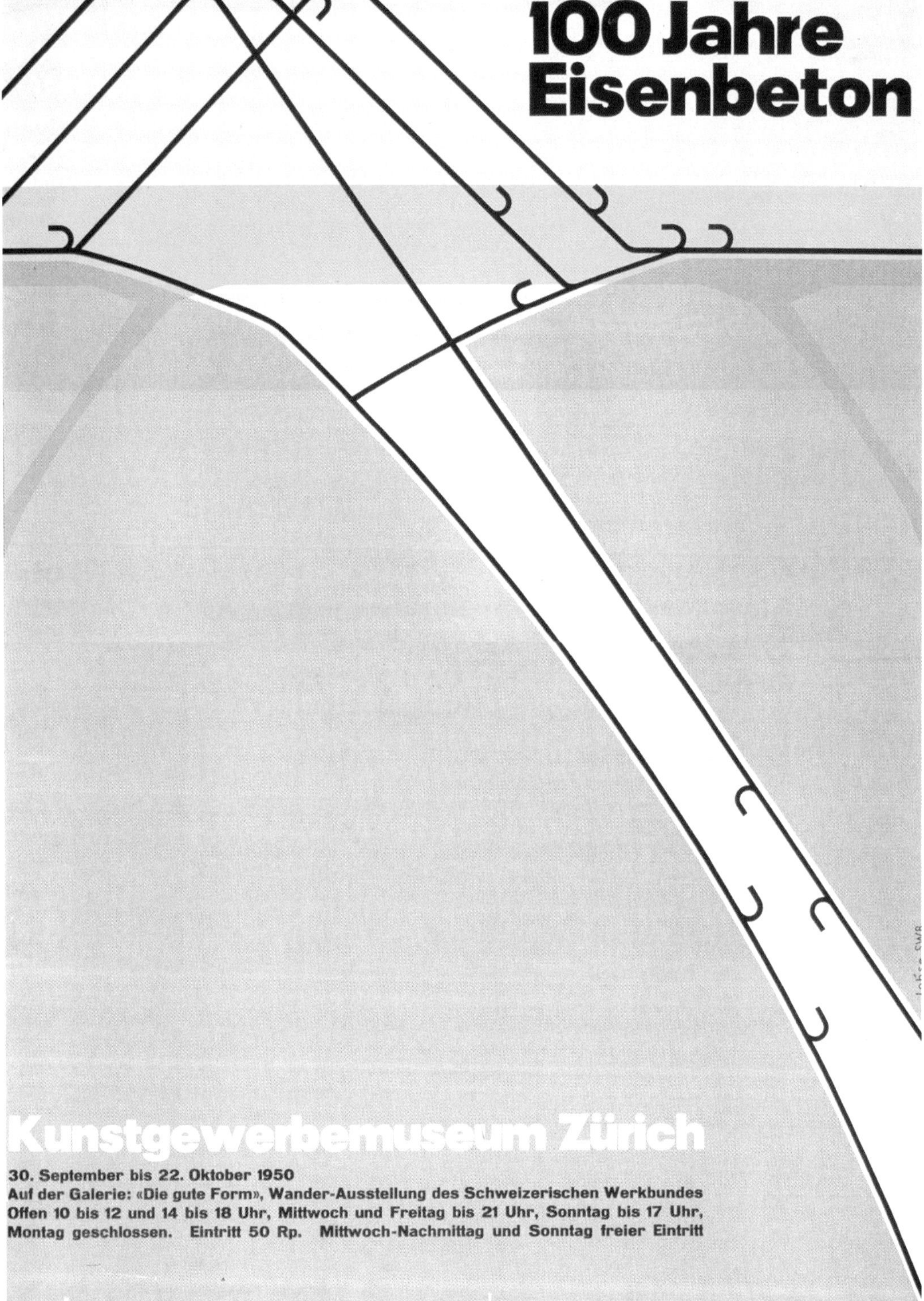

26

Richard Paul Lohse, “100 Years of Ferroconcrete”, exhibition poster, Zurich, 1950

Richard Paul Lohse was trained as a graphic designer and painter within the constructivist school and created the visual look for the **architecture magazine "Bauen + Wohnen"** established in 1947, from **1948** onwards. He was also on the editorial staff, and as a contributing writer he composed a long series of articles until leaving the magazine in 1956. "Bauen + Wohnen" strongly advocated the industrialization of architecture, it was a publication that did not concern itself with the linear forward progression of "building art", but focused rather on the then current question of reconstruction and of

Book *Neue Ausstellungsgestaltung*, edited and designed by Richard Paul Lohse, 1953

a new orientation towards planning. By its third year, the magazine was already being published in three languages: not only German, but also French, and English. Over the course of but a few editions, Lohse developed a typographical instrument that lasted far beyond 1956 and remained well into the 1970s (In 1980 "Bauen+Wohnen" merged with "Werk").

——— The magazine covers refer to the contents, whereby Lohse employed his concept of field grid graphic design. Displaying the table of contents on the cover page as well as inside the magazine was an innovation. The two-toned large 28

Bauen + Wohnen
Title page 8/1950 showing a chair by Willy Guhl

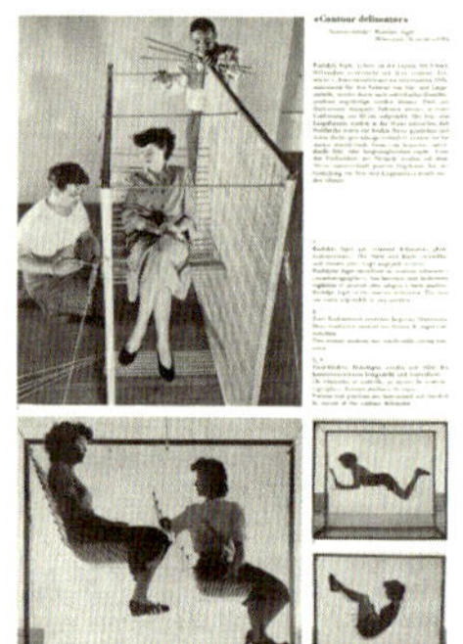

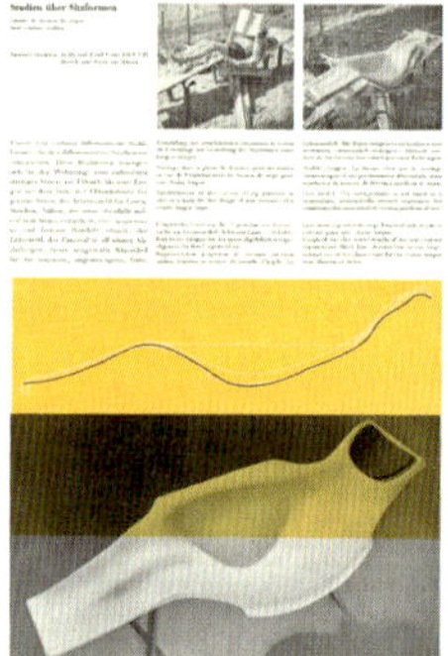

Double page spread 8/1950

edition numbers were placed near the binding. The multilingualism was signified by a small title “Construction + Habitation” or “Building + Home” on either the far bottom or top border of the cover, and the Z-shaped interconnection engendered a powerfully charged group of elements. Lohse’s **poster “100 Years of Ferroconcrete”** produced in 1950, with its x-raylike design turns a concealed building sheathing (the image of an aqueduct located in le Châtelard, in the Swiss canton of Valais, created by the engineer Robert Maillart), into a defining element of style and a definite eye-catcher.

Title page 9/1950

Title page 6/1954

In 1947, designer and interior architect **Jacob Müller** designed this set consisting of a small table and two stools, which he named **"Plio"** (cf. French "plier"). Plio is an ingenious piece of concentrated resourcefulness. First, the **table**: Its top is finished with birch veneer; The beechwood legsare coupled in pairs at their upper ends, and their geometry allows them to be laid flat on top of each other. The apron is reduced to two slats on the ends, and the two **stools** can be stowed between them when folded. The stools: Their legs, too, are coupled, so that the stool can be folded flat. The seat itself is very thin, approximately 2 mm, and consists of highly flexible airplane plywood made of the finest veneers. The middle of the seat is fastened to the crosspiece, the sides are moveable. In this way, the seat's own tension is used to fix the feet; pegs

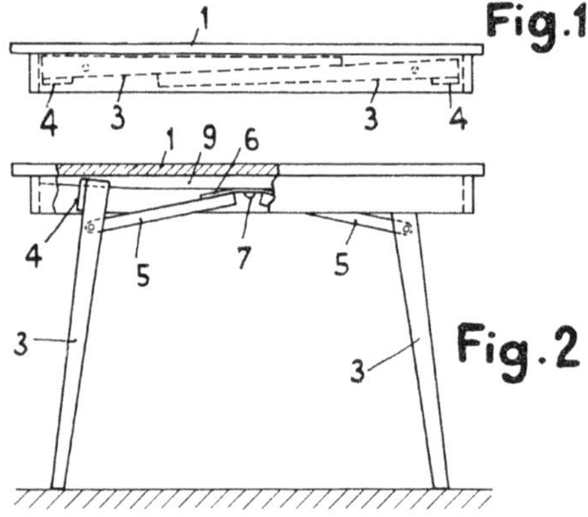

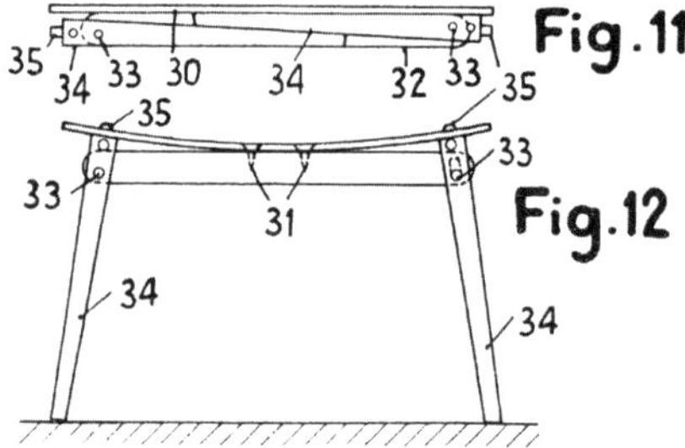

Plio, excerpt from the patent specification, 1948

at the top of each leg snap into the round holes in the seat. When the stools are folded, they slide into grooves in the table's end-slats, the table legs are. closed around them, and the stools are latched with two wire clamps There is a compelling logic to the whole, in which all the parts interlock and in which all are active and none passive. The sophistication of the: design is due in part to its consideration of not just two, but three states folded, unfolded, and the transition from one to the other – as in tying or untying a parcel, though without the aid of string or scissors. ——— Plio is a contribution to the theme "package furniture", which was intended to play an important role in rebuilding Europe after 1945. The **Aermo** company (Altdorfer-Ehrlich-Möbel) in Zurich carried Plio until about 1955.

Plio furniture set, table and two stools

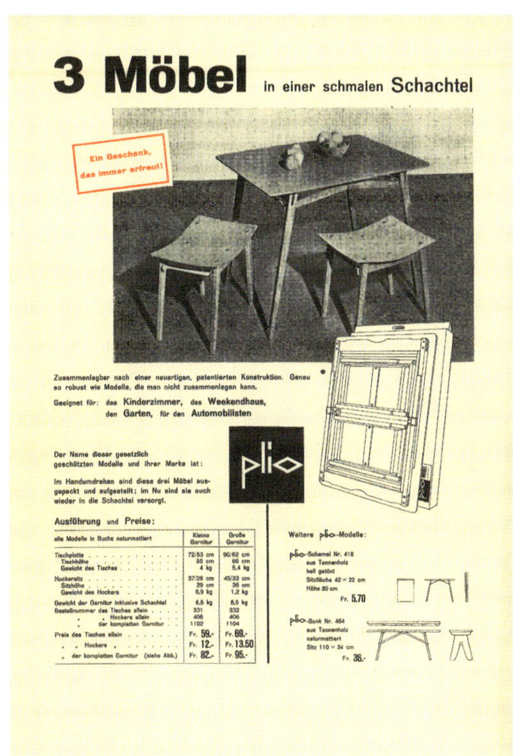

3 Möbel in einer schmalen Schachtel

Ein Geschenk, das immer erfreut!

Zusammenlegbar nach einer neuartigen, patentierten Konstruktion. Genau so robust wie Modelle, die man nicht zusammenlegen kann.

Geeignet für: das Kinderzimmer, das Weekendhaus, den Garten, für den Automobilisten

Der Name dieser gesetzlich geschützten Modelle und ihrer Marke ist:

plio

Im Handumdrehen sind diese drei Möbel ausgepackt und aufgestellt; im Nu sind sie auch wieder in die Schachtel versorgt.

Ausführung und **Preise:**

alle Modelle in Buche naturmattiert	Kleine Garnitur	Große Garnitur
Tischplatte	72/53 cm	90/62 cm
Tischhöhe	55 cm	66 cm
Gewicht des Tisches	4 kg	5,4 kg
Hockersitz	37/26 cm	45/33 cm
Sitzhöhe	29 cm	36 cm
Gewicht des Hockers	0,9 kg	1,2 kg
Gewicht der Garnitur inklusive Schachtel	6,5 kg	8,5 kg
Bestellnummer des Tisches allein	331	332
" " Hockers allein	406	406
" der kompletten Garnitur	1102	1104
Preis des Tisches allein	Fr. 59.-	Fr. 69.-
" " Hockers "	Fr. 12.-	Fr. 13.50
" der kompletten Garnitur (siehe Abb.)	Fr. 82.-	Fr. 95.-

Weitere plio-Modelle:

plio-Schemel Nr. 418
aus Tannenholz
hell getönt
Sitzfläche 42 × 22 cm
Höhe 30 cm
Fr. 5.70

plio-Bank Nr. 464
aus Tannenholz
naturmattiert
Sitz 110 × 34 cm
Fr. 36.-

Catalogue page

Since the 19th century, our internal image of a **factory** is that of a voluminous building with a saw-tooth roof. However, this is only the concept of a principle, not of a shape; no specific idea about the construction, such as the length-width-height ration, can be inferred from it. ——— In **1954**, for the project of developing an elastic band production hall for the textile industry in **Gossau**, the young architect **Heinrich Danzeisen**, partner in the office of Danzeisen 32

und Voser (St. Gallen), had a vision of structure and shape. The engineer **Heinz Hossdorf**, whose contribution was decisive in the realization of this project, later called the idea ingenious. Danzeisen's concept consisted of a series of slanted cylindrical vaults spanning 28 meters to arch over the hall. The inclination between the vaults created crescent-shaped openings, which allow daylight into the production hall. ——— It was a big disappointment when the

static calculation reveled that the necessary vault thickness had to be 40 cm. This contradicted Danzeisen's striking idea and was, furthermore, financially unrealistic. Both budget and building were trimmed when Hossdorf suggested to solve the problem by using the crescent-shaped openings as a static height. Only 30 years old at the time, Hossdorf achieved a composite action through an innovative synergetic combination of concrete vaults and steel framework. 34

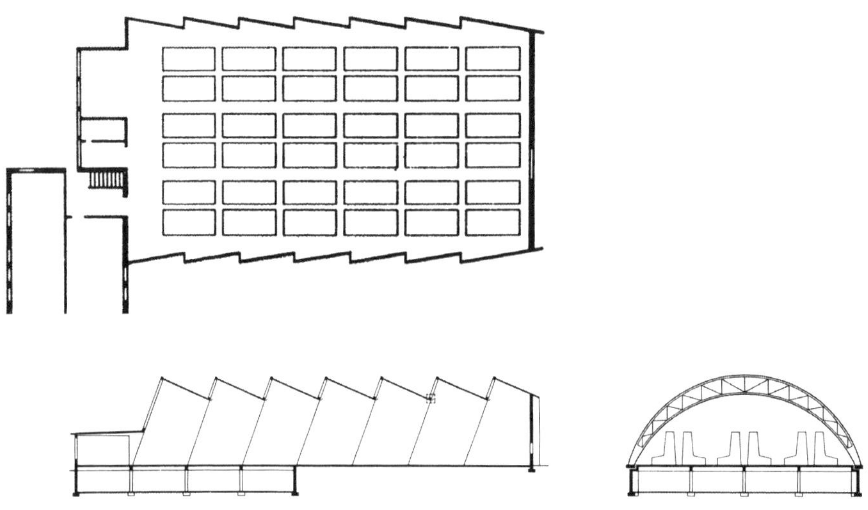

______ The seven vaults were jetcreted in pulsing procedure, meaning that complex scaffolding for two vaults was built and then taken down from the front and reassembled in the back. While removing the scaffolding, the composite action for each set of two vaults had to secured. Danzeisen and Hossdorf had only praise for the local constructors, who had successfully taken up the challenge of producing a
35 demanding, highly precise and reusable scaffolding.

32/33 Elastic band factory in Gossau, photo 1955
Floor plan, longitudinal section, transverse section
Photograph of the construction site, 1954
Current condition, 2005
Production hall, view to the north, 1955

View at night, 1955

37

Max Bill appeared on the architectural scene in 1933 with his own residence and studio in Zurich. He received international attention for his groundbreaking design for the Swiss Pavilion at the 1936 Milan Triennale. The **Villiger House**, built in **Bremgarten** in **1942**, was Bill's second residential design. The range of materials available to the architect was restricted by the wartime economy; government controls on cement and iron ruled out his using concrete or steel. Bill, who made the response to externally or self-imposed constraints a constant theme in his work, designed a wood house. In a programmatic contrast to the Swiss *Heimatstil* (let alone its National-Socialist counterpart in Germany), he employed wood straightforwardly as a building material, unburdened by any ideological agenda. ——— The house, which unfortunately no longer exists, took the form of an L-shaped pavilion set on a nearly square ground plan. It comprised six rooms and a garden shed under a common flat roof. The exterior walls consisted of Durisol panels 150 cm long by 50 cm high, with the members of the wood-

The Villiger House shortly after completion

frame bearing structure spaced accordingly. In their formal expression, the walls were subordinate to the flat roof plane. Free-standing columns, also spaced at 150 cm, supported the edge beam of the roof structure, which extended beyond the walls facing the terrace to form a covered, wood-floored ambulatory. As post footings Bill used boulders dug out of the terrain during the excavation of the site. Because it rose above the land rather than clinging to it, the house stood, as it were, in an immanent discourse with the prevailing ideological positions of its time. The roof was the dominant feature in determining the house's form. It was as if the architect had reversed and, in so doing, subverted the claim of the anti-modernists' *Blut und Boden* faction. Even as he took up wood as a traditional material, Bill broke with European, not to mention Alemannic, building practice, and in fact consciously looked to Japanese architecture as a model. Its use of wood in this way makes the Villiger House a particularly lucid affirmation of ar-
39 tistic and intellectual freedom.

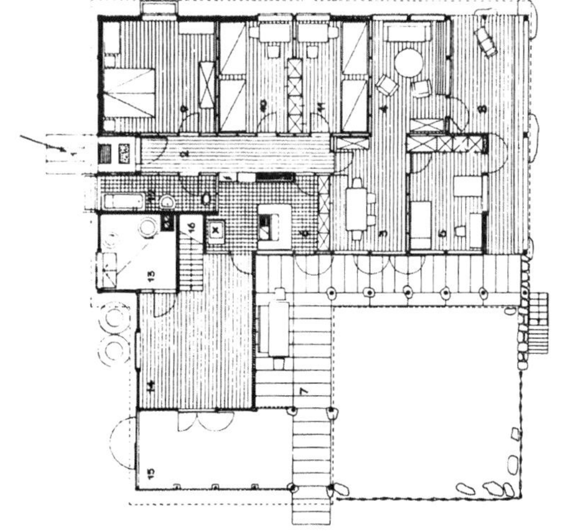

By 1931 **Max Bill** had already received international recognition as a graphic artist. At that time, in the book “Gefesselter Blick” (“The View Unbound”, published by Heinz and Bodo Rasch) he incidentally mentioned the idea of a so-called compact letter, in which an A4 sheet of paper could be written on and then folded and glued to serve as its own envelope, which later became known as the “Aerogram”. His most important works as a poster designer were created in the 1930s. ——— The example documented here from his later works, the **exhibition poster** for a **1977** retrospective featuring **“New Design in Zurich”** at the **Arts and Crafts Museum** in **Zurich**, in its concise construction is loosely reminiscent of his earlier works. Bill did not like to look backwards; he aligned himself with avant-garde forward-looking views. Is the Zurich Coat of Arms hinted at in the typographical diagonals utilized in this poster? (The Coat of Arms pattern, however, runs from the upper left to the lower right). This remains unclear, as it should. In many of his works, Bill gives the impression that they are cursorily implemented ideas. But soon it is clear that this rashness does not lack reflection. Bill’s **1952** book on the topic of **Form**, subtitled “A balance Sheet of Mid-Twentieth-Century Trends in Design” is a fascinating selection of objects, some strikingly beautiful and some totally utilitarian, from the Alpine dairy ladle to high-voltage masts. It is also a document that expresses his free-spiritedness. 40

Eine Bilanz über die Formentwicklung um die Mitte des XX. Jahrhunderts
A Balance Sheet of Mid-Twentieth-Century Trends in Design
Un Bilan de l'Evolution de la Forme au Milieu du XXe Siècle

Max Bill

FORM

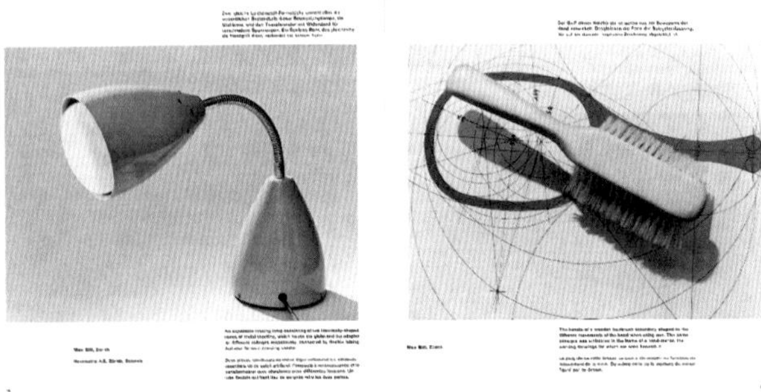

Max Bill, Book *Form. Eine Bilanz des Formschaffens um die Mitte des 20. Jh.,* 1952. Dust jacket and double page spread of two products designed by Bill.

kunstgewerbemuseum zürich

um 1930

in zürich

neues denken neues wohnen neues bauen

3. sept. — 6. nov. 1977

di — fr 10 — 18
mi — 21
sa + so 10 — 12 + 14 — 17
mo geschlossen

bill
77

Max Bill, "Zürich um 1930", exhibition poster, Zurich, 1977

MÖBEL
MEILI

The **railway platform roofs** of the suburban station in **Winterthur-Grüze** manifest an underlying ambition of their designer, **Hans Hilfiker.** Hilfiker, an electrical engineer, was responsible for stationary facilities at the Swiss Federal Railways. He possessed a prescient, and still current, understanding of industrial design as the comprehensive integration of all its constituent parts. In a 1952 essay, he put forth a new concept for the form and function of roof structures for station platforms ("Über Funktion und Form des Bahnsteigdaches"), which was realized in his design for Winterthur-Grüze in **1955.** Hilfiker's fundamental analysis led to a thorough reformulation of the platform roof that effectively posited a new type. Winterthur-Grüze was an important example of the notion of prefabrication. It remained a prototype, however, as few new railway stations were built during this period and there was insufficient pressure to replace the existing platform shelters. ——— Hilfiker criticized two aspects of the era's existing platform 44

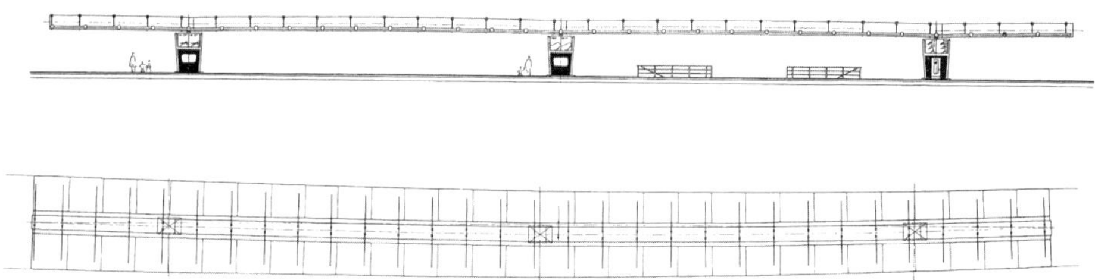

42/43 The platform roofs after completion of the first of the two roofs, 1955

Prefabrication and delivery of the pipe girder sections

Assembly at night, 1955

Elevation and floor plan

46/47 Connecting two sections of the central pipe girder

roofs: the great number of columns, which hindered movement on the platforms; and the unsystematic placement of elements that had become part of the railway and its functions over time: *operational* equipment like conduits, signals, communications devices, and electric consoles; *public* amenities such as signs, timetable displays, clocks, mailboxes, and telephone booths; and *commercial* accessories, including vending machines, poster frames, display windows, and advertising space. He proposed small structures that would at once support the roof construction and, with their portal-like form, accommodate these installations in a kind of modular building kit. ——— Instead of using traditional roof construction (transverse beams on columns, purlins, roof framing, decking), Hilfiker suspended broad roof panels from either side of a central pipe girder. The central girder runs 33 m between piers – three times the structural span found in conventional platform shelters of the time. With a diameter of 120 cm, the girder

is supported at just three points over its 90-m length and cantilevers a full 12 m beyond the end piers. The roof itself consists of prefabricated, steel-reinforced ceramic slabs stayed by cables affixed to the top of the girder. The slabs are conceived as large interlocking tiles, each with an area of 10 m, and shed rainwater inward toward the central girder. At non-loaded points, they are barely 3 cm thick. An uninterrupted band of fluorescent light tubes accentuates and "contours" the edge of the roof. ——— Precise logistical planning allowed the structure to be built very rapidly. A special train transported the pipe sections to the site, where they were lifted into place during the night. The suspension of the roof elements was carried out at a similar pace. Winterthur-Grüze brings together architecture, design, railway technology, public service, and logistics in an exemplary case of integrated thinking, striving towards a goal which combines technical and operational aspects as well as public convenience.

We take the liberty of overstepping the temporal framework of our selection: **Wilhelm Kienzle**'s **cactus watering can** dates back to **1938**. The Blattmann hardware factory in Wädenswil produced it for decades. The design takes into account that a windowsill usually offers little room. In contrast to watering cans for ornamental plants, the spout does not project forward, but upward from the top of the watering can. The body possesses an elliptic base. All the formal elements result from the basic goal of saving space. The tall and slender handle is shaped so that the little watering can can be tipped from the wrist; the concave top of the water container serves to limit the volume of water so that it pours only from the spout. The inlet and the shaping of the handle permit easy filling; the entire configuration is designed to let air flow in as the water pours out so that the water does not slosh and spill. The tip of the spout is painted yellow so that it is easy to see between leaves even for people with poor eyesight; the comb-like element at the top of the spout can be held with one's fingers to direct the water precisely. This is our praise for the object's rationality in its entirety and in detail. Above and beyond rational considerations, the object's appearance is magically charming. Designer Wilhelm Kienzle himself wrote that it looked "like a bird sitting in the bushes." 48

In around **1955**, designer **Benedikt Rohner** developed a **wooden plan cabinet** that appears more like home furniture than the usual metal cabinets, a form remaining current to this day. Rohner designed this piece for the cabinetmaker **Ph. Oswald** in Oberglatt near Zurich, a well-known manufacturer of sophisticated furniture, who has been supplying it continuously ever since. —— The decisive elements of this design are the drawer bottoms, extended towards the front, with two circular holes for opening the drawer. The bottom of each drawer protrudes to the sides, and the drawer glides in and out in grooves cut into the side walls; the inside of the drawers is in solid fir glued onto plywood. There are no ball bearing mounted fittings. The model for this design is not a piece of professional office furniture made of metal, but rather furniture for living quarters, for example a wardrobe with English drawers. —— The cabinet is available in two sizes: 108 x 84 x 72 cm, and the larger model, for example for posters, is 138 x 108 x 72 cm. The outer sides of the plan cabinet are finished in birch veneer, the drawer faces are set off by a black synthetic resin laminate.

It all started with a view of the Rhone valley from a height of 2,000 meters. This is where the shining river winds its way through the valley in the evening twilight and the shadows of the mountains become palpable. This is also where, in **Saflisch** (Simplon Massif), the young architects **Heidi and Peter Wenger** built themselves a wooden tent as a vacation home – the **"Trigon"** house (**1955, modified in 1976**). The house appears suspended in its own supporting structure. It consists of individually anchored triangles in five primary structural planes which total 10 m in length; the foundation is made of boulders which are given a smooth stucco finish. The cross-section of the main living space is an equilateral triangle whose sides are approximately 6 m long. Originally, the house was entered through a hatch; a narrow

Exterior and interior views: terrace/shutter

spiral staircase penetrated the floor and led to the gallery (The small room underneath the house was later needed to store firewood and is not shown in these photographs). The space above is big enough for a sleeping loft. ——— The house's main feature is the huge front window shutter which turns into a terrace and vice versa. Could there be a more attractive architectural interpretation of arriving, being there, and departing? ——— The dimensions of the house are very pleasant; the feeling is not claustrophobic, but open. The shutter / terrace, with its 5.5 m long sides, produces a surface area of approximately 15 square meters and weighs 400 kg. It was initially raised and lowered by means of a winch. When electricity became available, the winch was replaced by an electric motor. The terrace is a triangular

space between heaven and earth, surrounded by larch trees, through the branches of which the town of Brig can be seen 1,500 m below. On the side of the house facing the mountain, two hinged panels analogous to the shutter / terrace were added in 1976: the tip of the triangle for the sleeping loft and a trapezoid for the living area. Furthermore, the house was also equipped with triangular windows which can be rotated around their central vertical axes. Originally, the elevation facing the valley contained rectangular French doors, the side facing the mountain had almost no openings. The 1976 remodeling greatly enhanced the house's clear statement by making the longitudinal axis more visible, emphasizing the triangles in the front and rear elevations, and adding the spherical kitchen. 52

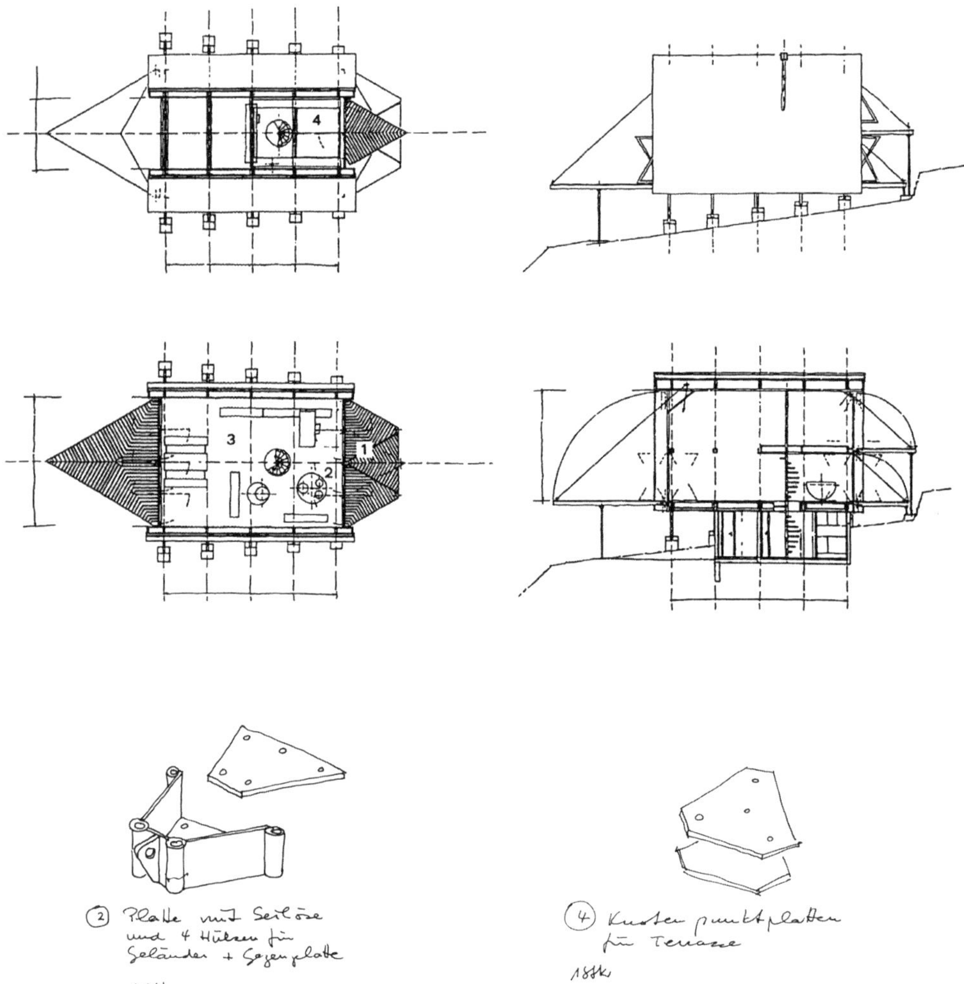

Construction site in the summer of 1955
showing Peter Wenger (center) and Heidi Wenger (right)

Floor plans, section, elevation, sketches of details.

54/55 Photograph, c. 1976

The Baltensweiler company began after the couple **Rosmarie und Rico Baltensweiler** (she an interior designer, he an electrical engineer and technician) had designed a flexible **floor lamp** in **1951**. He worked out the technical details meticulously, while she determined the stylistic idiom. It was a private act of liberation from the misery of the conventional floor lamps of the day with a turned wooden shaft and a rigidly attached shade of parchment or fabric, and came just as the "Wohnbedarf" furniture store was preparing to launch a related alternative, the Akari lamp. Publication of the Baltensweilers' unique piece in the progressive magazine "Bauen + Wohnen" sparked interest in the design and gave the couple the impetus to found their own studio in **Ebikon** and to mass-produce the lamp under the name **Modell 600**. Its three joints and rod-like construction give it a more technical appearance than the Akari lamps by Isamu Noguchi, but it has a grace all its own – not least because of the circular co-reflector fastened directly to the light bulb with a wire clump. This creates an effect reminiscent of a cosmic

event such as a solar eclipse. The design combines the principle of balance (cast iron base and counterweight on the boom) and a swivel joint in which an interior tension spring creates a stablizing frictional resistance. ——— The aluminum reflector, nicknamed “Chinese hat” by the makers, had a conic neck in the early years. Jacques Tati used this model of the lamp in his film “Mon oncle” (1957). In 1964, the company went to a two-part reflector with a cylindrical tube containing a press-button switch. This tube as well as other parts were commercially avaiable semifinished products. Keeping design and production under one roof – a principle Baltensweiler still follows – the company produces a line of related models in which even the inconspicuous details are elegantly resolved. The Baltensweiler company is among the industry pioneers in using new illuminants: 1973 halogen lamp (Halo-250), 1984 fluorescent lamp (Manhattan), 1987 compact fluorescent lamp (Aladin), 2003 LED (Modell Zett, electric power consumption
 4 watts).

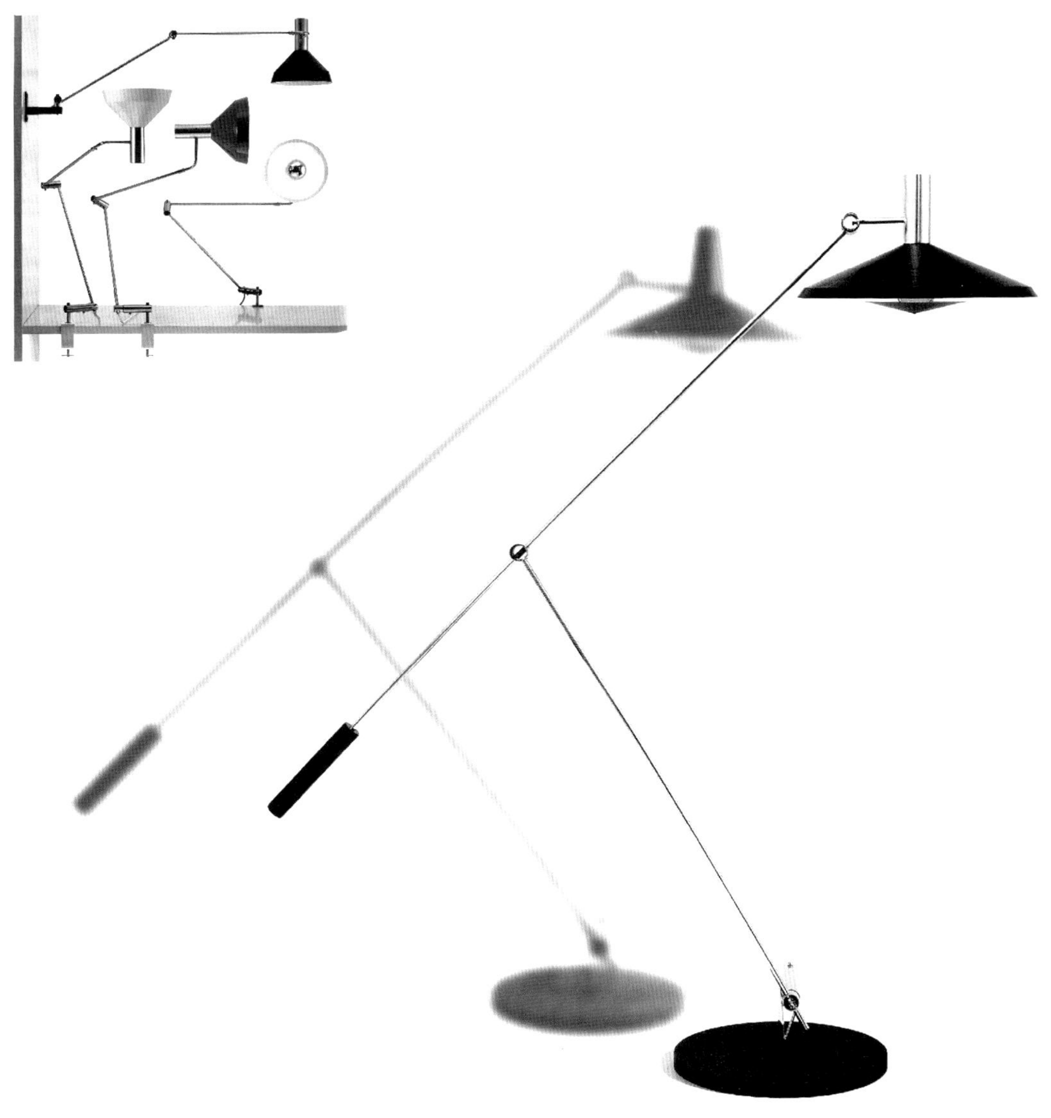

The Swiss version of the **Trans Europ Express (TEE)** was a distant descendant of the category "luxury trains" and reminds one of the heyday of railroad travel. The concept of the Europe-wide network of TEE trains, agreed on in 1954, followed from the founding of the EEC (European Economic Community). Despite the fact that Switzerland did not belong to the EEC, the country was included in the Western European railroad concept because of its geographical location. The goal of the TEE network was to keep well-heeled passengers from choosing airline travel over the train. The Swiss Federal Railways (SBB) commissioned the SIG (Schweizerische Industriegesellschaft) in Neuhausen with the design of a Swiss TEE, which, in contrast to the diesel-powered German and Dutch ones, was electrified. The locomotives of the Swiss trains were even constructed for four different European electricity systems, the first such design ever. This "hidden" technical achievement represents an entire subterranean world of Swiss know-how. But the train also had a very visible and effective public appeal. ——— **Walter Henne**, an architect from the town of Schaffhausen, was commissioned to design the train in **1959**. Schaffhausen and Neuhausen are neighboring towns; in those days, design still took place at the plant itself.

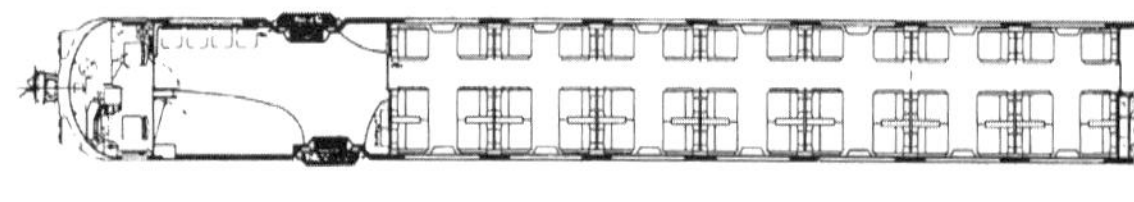

SBB's TEE train, design by Walter Henne and SIG, 1961

Shell of the train, c. 1960

Floor plan of the train (detail)

But above and beyond being in a convenient locality, Henne was a very capable designer who had great sensibility for lines and for the effects of materials. He had previously been responsible for designing the SBB's "standard cars" at SIG. The TEE boasted passageways without doors for moving from one car to the next, a sophisticated joint mechanism which was premiered in Europe on this TEE train. ——— The train, whose total length was 160 m, had 168 seats, all of them first class. The restaurant with a separate bar was an important meeting place for sophisticated travelers. Ladies' and gentlemen's restrooms were separate, and the ladies' featured a vanity. Cars were air-conditioned, and there were electrically powered Venetian blinds with pale yellow slats between the sealed double-glazed windows. The comfortable seats were fashioned after international models (Italy, the US – clipper class airplane seats). In fact, the distances covered were almost too short for this level of comfort. The train was deployed on routes such as Zurich – Milano or Milano – Paris. Inaugurated in 1961, it ran during the day and had no sleeping car. The TEE carries the attention to detail and the craftsman's pride characteristic of the 1950s into the cooler 1960s. (Continued on page 142)

Dining car with bar

The well-known engineer **Mirko Roš** engaged **Max Bill** as his collaborator in the design competition for the **highway bridge** at **Tamins** (between Reichenau and Flims). During World War II, Bill had intensively studied the work of Robert Maillart and in 1948 had published the first and very commendable monograph of this great maverick. ——— Bill persuaded Ros to depart from his original scheme to build an arched bridge and to adopt his vision for the design. He conceived the roadway as a box girder supported by a strut frame, the diagonal members of which soared to meet the roadway girder close to its mid-point. Bill termed the middle section of the roadway – the section spanning the gap between these struts – the "lengthened vertex" of the box girder construction. The lower section of the two diagonal struts was forked, so as to broaden the base as well as allow the accompanying vertical column to stand between the paired feet and thus transfer its load directly into the foundations. The proposal won first prize in the competition and was carried out in **1966**, albeit with alterations. Bill lamented that the engineers (the firm Anschwanden & Speck, who took over the commission following Roš' death) lacked the courage to employ straight members for the angled struts as called for in the original plans, instead using slightly arced segments to achieve the effect of an arch. ——— This is a striking case in which an avowed designer had a decisive influence on such a highly specialized field as bridge engineering. Bill's impulse to reduce a problem to its essentials, which arose from his strong self-conception as a generalist, posed a challenge to the specialist discipline, and his belief in a simple solution ran counter to the usual development dynamic of complex problem-solving processes. At the same time, however, it is an expression of his conviction that the designer must search out the characteristics and behavior of the materials and give them plausible form.

Vanity in the ladies' restroom, 1961

Max Bill / Mirko Roš (with Aschwanden & Speck, engineers): Lavinatobel Bridge near Tamins, 1966

The visitors of the international **garden show (G59)** in Zurich could find everywhere a seat for outdoor use designed by **Willy Guhl** in **1954**. For years, the manufacturer, Eternit (Niederurnen), had been well on the way to major international success with its corrugated panels. ——— An artisan whose pride derived from his woodworking skills, Guhl discovered his creative interest in plastic seats and free forms as early as 1947. In this case, he created a structure shaped like a ribbon without beginning or end, a low rocking chair that would not sink into the lawn because of its large bottom surface. The Eternit loop was an apt expression of the times. No seat in Europe had ever been so low.

Even the name Eternit expressed the material's claim of eternal durability; weatherproof and frostproof, barely hygroscopic – needing almost no maintenance, but sophisticated in form. Eternit still (and again) carries this easy chair. Since fiber cement must be asbestos-free today and therefore has different structural properties, Guhl redesigned the chair and incorporated two supporting ribs for reinforcement. ——— This loop's swinging free form reminds one of the late 1950s, when a growing segment of the population could enjoy vacations, trips abroad, private cars, and better-furnished and decorated homes thanks to the beginning economic boom.

The "Loop Chair" by Willy Guhl, 1954

Scobalit plastic chairs by Willy Guhl, 1951

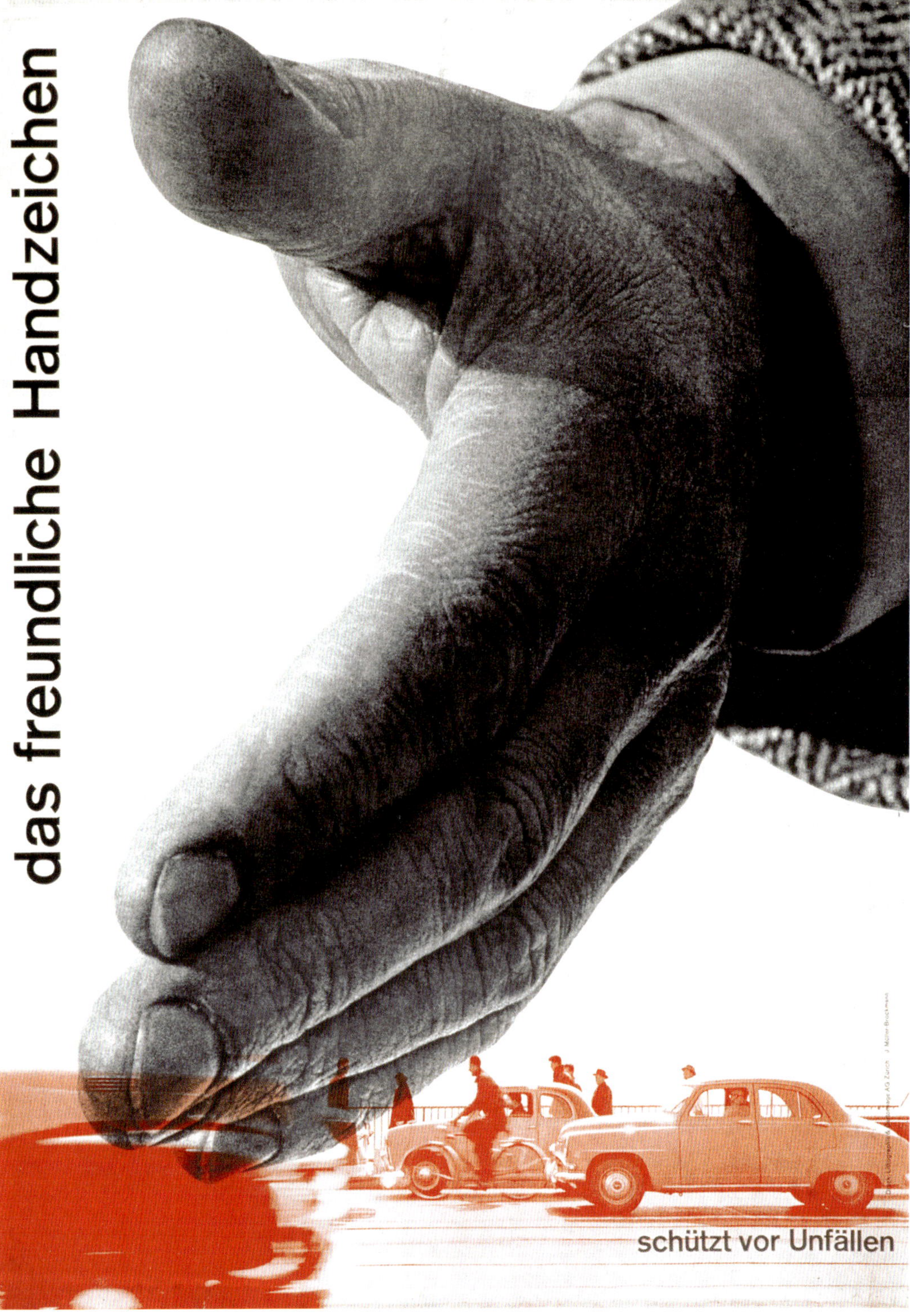

Joseph Müller-Brockmann, poster on traffic education, 1955

“The renunciation of a free, subjective style of representation opens up space for composing forms relevant to a subject in a tension-packed way [...] The more anonymous and objective a graphic element is, the better it is fit to convey an idea.” **Josef Müller-Brockmann**, who wrote this so clearly in **1961**, took the step from experienced illustrator to representative of objective graphic design around 1950. His work and internationally acclaimed books did much to boost the fame of Swiss graphic design. His central design question is the investigation of the topic, which he believes to be the “essence” of graphic design. What is essential? One thing is clear to him: It can not be captured by a mere depiction. He therefore puts something else in the place of the usual genre image: the image as “implementation”. He believes (black and white) photography, as opposed to drawing, to be the more objective medium. He does not see photography as “documentary”, however, but uses it to create pictures of his own imagination. The photographic model is already a goal he has set his sights upon, and something he discusses with the photographer. He uses the large repertoire of flexible components at his disposal to accomplish his aims: blurred depth of field in the front, motion blurring, different intensities of light, multiple exposures, differences in size, overprinted colors...–these all help give form to the idea to be communicated using technical means, means chosen by the designer, who must search for the correct method to convey the message at hand. After 1955, he transmits linguistic information almost exclusively using the typeface Akzidenz-Grotesk in markedly differing sizes. Müller-Brockmann did not invent this, he learned from the 1920s. After 1945, the period between the world wars was an exciting rediscovery throughout the (Western) world. Objective graphic design must transcend objective distance. It is not an approach which can simply be “applied”, because the search for the “essence” eludes all routine. The worth of Müller-Brockmann’s work lies in the strength of his solutions.

Tonhalle Grosser Saal 11. Volkskonzert Dienstag, den 6. Januar 1953 20.15 Uhr

Leitung **Erich Schmid**

Solisten **Elsa Cavelti** Alt

Heinz Rehfuss Bass

Introitus, für Streichorchester **R. Oboussier**

Sinfonie Nr. 7 (delle canzoni) **F. Malipiero**

Herzog Blaubarts Burg **B. Bartók**

Karten zu Fr. 1.-, 2.- und 3.- (inkl. Steuer) Vorverkauf Tonhallekasse, Hug & Co., Pianohaus Jecklin, Reisebureau Kuoni

Joseph Müller-Brockmann, concert poster for Tonhalle Zurich, 1953

67

1. orchesterkonzert

mittwoch, den 4. juni 1969
leitung/ erich leinsdorf
solist/ isaac stern, violine
c. m. v. weber/ freischütz-ouvertüre
l. van beethoven/ violinkonzert in d-dur, op. 61
igor strawinsky/ le sacre du printemps

extrakonzert

sonntag, den 8. juni 1969
isaac stern, violine
alexander zakin, klavier
werke von bach
brahms
prokofieff
bartok

2. orchesterkonzert

dienstag, den 10. juni 1969
leitung/ antal dorati
solist/ claudio arrau, klavier
joseph haydn/ sinfonie in b-dur, nr. 98
richard strauss/ till eulenspiegels lustige streiche, op. 28
johannes brahms/ klavierkonzert in d-moll, op. 15

musica viva-konzert

dienstag, den 12. juni 1969
duo alfons und aloys kontarsky, klavier
christoph caskel, schlagzeug
werke von bernd a. zimmermann
earl brown
karlheinz stockhausen
pierre boulez

konzerte
junifestwochen 1969

tonhalle —
gesellschaft
zürich

3. orchesterkonzert

dienstag, den 17. juni 1969
leitung/ rudolf kempe
solist/ zino francescatti, violine
karl amadeus hartmann/
kammerkonzert für klarinette, streichquartett
und streichorchester (uraufführung)
felix mendelssohn/ violinkonzert in e-moll, op. 64
l. van beethoven/ siebente sinfonie in a-dur, op. 92

2. extrakonzert

donnerstag, den 19. juni 1969
arturo benedetti michelangeli
werke von clementi
schumann
ravel

4. orchesterkonzert

dienstag, den 24. juni 1969
leitung/ wolfgang sawallisch
solist/ arthur rubinstein, klavier
arthur honegger/ monopartita
peter tschaikowsky/ klavierkonzert in b-moll, op. 23
robert schumann/ zweite sinfonie in c-dur, op. 61

5. orchesterkonzert

dienstag, den 1. juli 1969
leitung/ rudolf kempe
solisten/ christa ludwig, alt
waldemar kmentt, tenor
w. a. mozart/ sinfonie in b-dur, kv 319
gustav mahler/ das lied von der erde

vorverkauf

tonhallekasse
musikhaus hug
pianohaus jecklin
reisebureau kuoni
filiale oerlikon kreditanstalt

preise

fr. 10.- bis 35.- orchesterkonzerte
fr. 10.- bis 30.- extrakonzerte
fr. 5.- bis 11.- musica viva-konzert

entwurf
josef müller-brockmann
druck
bollmann zürich

Joseph Müller-Brockmann, concert poster for Tonhalle Zurich, 1969

68 Joseph Müller-Brockmann, advertising, Zurich, c. 1955

BALLY
Schuhe
TAXI
Tel. 23 56 00

1962–1973
STUDIES AND CONCRETE CASES

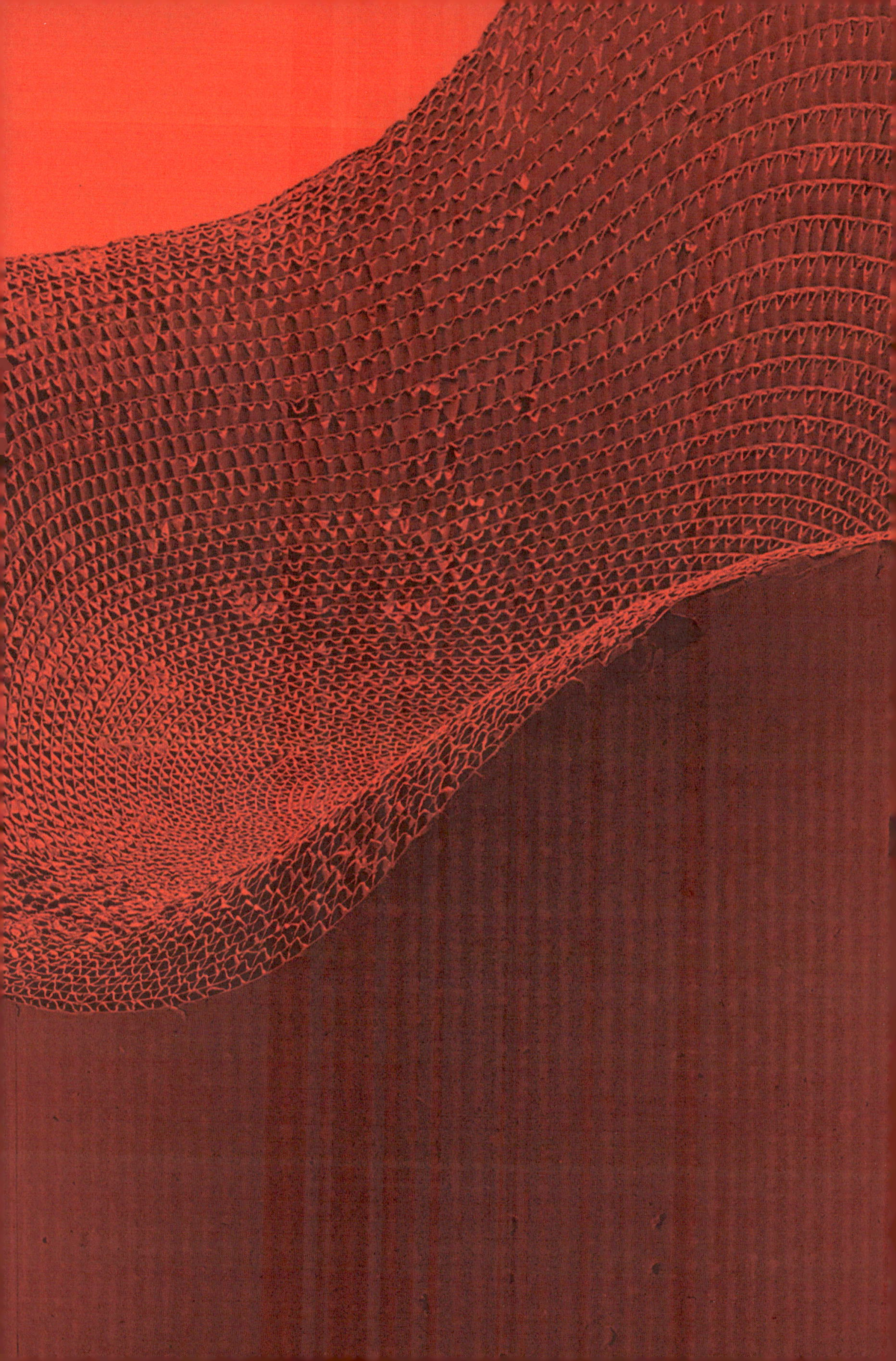

The beginning and end of the second section can be defined fairly precisely. The beginning: The construction of the Berlin Wall in August of 1961 is a drastic event that brings the Cold War to everyone's living room. The shadows of world politics increasingly darken everyday life. For the time being, Switzerland's innocence of the 1950s vis-à-vis world politics is over. A famine in sub-Saharan Africa (Biafra), the Cuban missile crisis in 1962, John F. Kennedy's murder in 1963, Brezhnev's deposition of the Soviet General Secretary Khrushchev in 1964, the beginning of the Vietnam War in 1964 are decisive events. The Swiss National Exhibition in Lausanne – Expo 64 – shows a dual image of Switzerland as a politically and socially conservative country hesitantly opening up to new cultural movements. The wave of Beatlemania rolling in from Great Britain sweeps over Swiss youth, triggering a youth culture much stronger than that ten years before in the age of rock 'n' roll. Everyday life is more relaxed than in the 1950s. Society has reached cruising altitude, sensational events occur less often. People have become accustomed to the economic boom and its amenities. **Concrete concepts** take the place of the frequently generalizing designs of the 1950s. Many works stand for themselves without having to simultaneously realize a universally valid model. The vibrations of the time have become fainter, the specific density of time is decreasing. It is well-known that a booming economy never encourages deep thinking.

All the more astounding to see how many high-quality designs emerge under such circumstances. The Theater am Hechtplatz in Zurich is such a jewel, the Cinéma Le Paris in Geneva is another, the Meggen Church a third. All three are distinctly individual solutions for the problem to be solved, and in all three, their individuality is attained by means that radiate supraindividuality (Two of these examples date back to before 1960, but, as is explained in the introduction, what characterizes a work as being typical for a particular era is not always congruent with the exact year of its inception). Some of the projects selected here do have their origins in the search for new models; for instance, the terrace housing in Umiken, which dates back to a concept from 1958, or Heinz Isler's groundbreaking self-supporting concrete domes, which were also developed before 1960, but did not remain merely prototypes, proving themselves to be very widely used structural solutions after 1960. In design, however, questions with totalitarian traits are posed only now – a remarkable time lag compared with architecture. The USM Haller modular office furniture system is a typical indication of the time for the catchword *infrastructure*, which makes its way from technical communication to everyday language during these years. In this age of economic overheating and affluent consumption, young students begin to doubt their parents' goals in life vehemently.

The dissent against prevailing societal conditions culminates and explodes violently in the May 1968 riots in Paris and other cities, including Zurich: a harbinger of postmodernism.

The end of these years of full employment is a hard landing: The oil price shock in the autumn of 1973 brings a quarter century of economic prosperity and long years of an economic boom to an end.

Armin Hofmann, exhibition poster “die gute Form”, Basel, 1954

Graphic designer **Armin Hoffmann**'s international appeal continues to be strong, particularly in the United States. An excellent teacher of many years standing (at the Kunstgewerbeschule (School of Arts and Crafts) Basel and of many courses in the USA), he is also an explorer; by constantly reflecting on his work and publishing these reflections, he stepped out of the daily work of teaching and became an internationally renowned authority. Unlike Josef Müller-Brockmann, who is several years older, he never founded his own agency and did very little work for private industry. This allowed him to develop his spectrum of expression more fully. Using his own idea of "Bildarbeit" (literally: image-work, a word he never used himself), he took advantage of the leeway his work in cultural promotion gave him, interrogating the possibilities of design without compromise. If he deemed it necessary – and he often did – to create a typeface especially for a given subject, he did so (e.g. the **poster** for the exhibition **"Tempel und Teehaus in Japan", Gewerbemuseum Basel**, **1955**). In this manner, he often prevents the public from routinely taking in the components of his posters – particularly the typographic elements – and sometimes deliberately compels them to scrutinize lettering as an elemental entity (with the inquisitiveness of first graders). Content, form and message of his work all rest on a design which lends the greatest importance to guided association. Unlike today's widespread understanding, however, "association" does not mean a vague affair of individual perception, but is something elicited by the design itself. His important works for Basel's chemical industry illustrate this belief (e.g. the packaging for an insecticide). ——— Hofmann's original approach has stood the test of time and is not bound to a particular stylistic idiom. As the designer of timeless posters, the author of books and other printed works, Hoffmann is still of strenuous interest to many of his remarkable young colleagues.

TEMPEL und
TEE-
HAUS
in JAPAN

Tempel und Teehaus
in Japan
Ausstellung im
Gewerbemuseum
Basel
täglich geöffnet
4. Mai bis 31. Mai
10-12 und 14-18 Uhr
Eintritt frei

Armin Hofmann, "Tempel und Teehaus in Japan," Exhibition poster, Basel, 1955

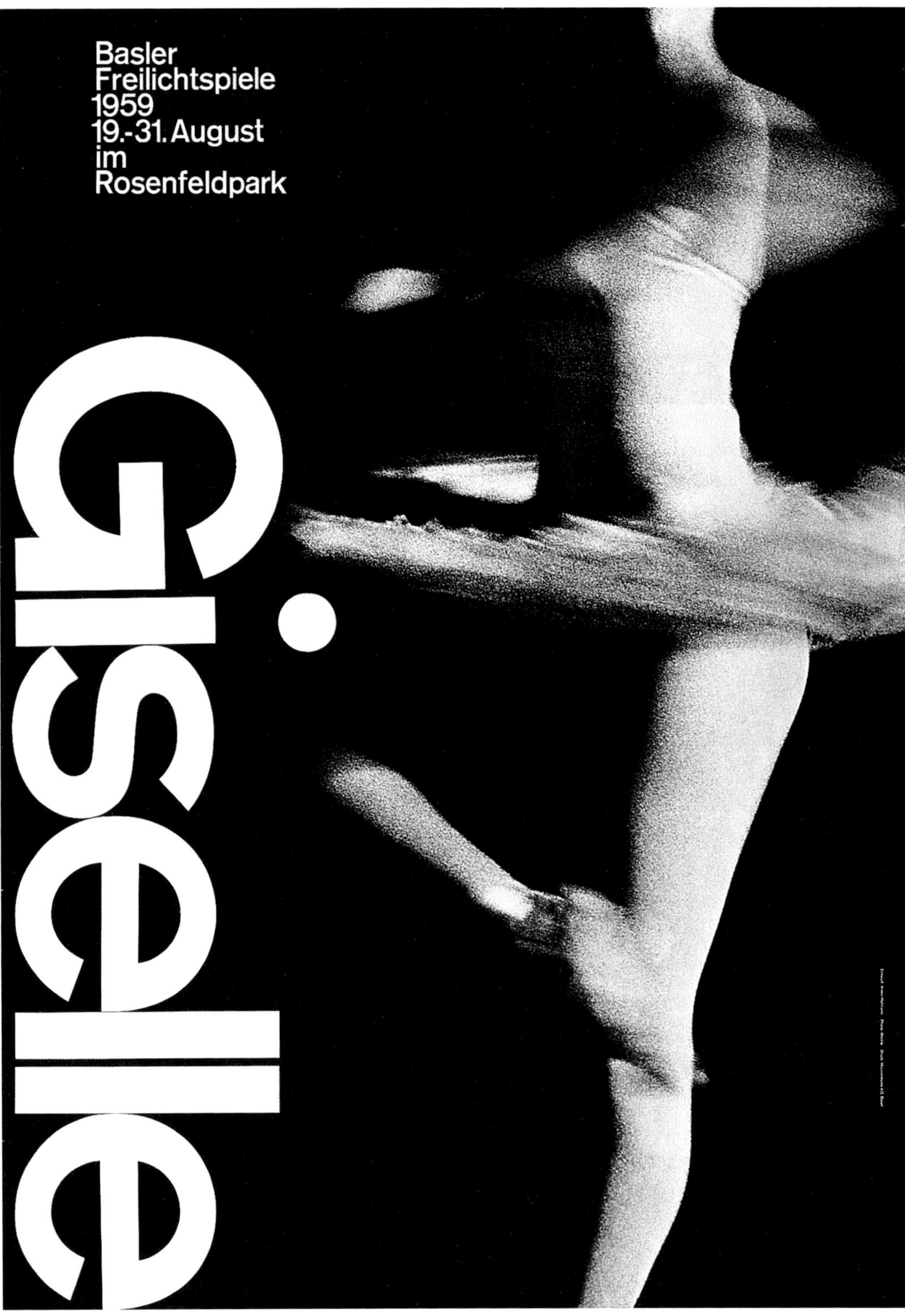

Armin Hofmann, “Giselle”, Poster for the Basler Freilichtspiele, 1959

Armin Hofmann, poster “Basler Theater”, 1968/69 season

Marc-Joseph Saugey from Geneva was an idiosyncratic amalgamation of identities: speculator, architect, playboy, and passionate chef. The Cinéma **Le Paris** in **Geneva**, opened in **1957** and today known as the **Auditorium Arditi**, reveals Saugey's architectural skill at the height of his powers. The cinema is one element of an angled multifunctional building complex Saugey designed, housing stores, offices, and apartments. The gable end, where the entrance to the cinema lies, runs towards the **Avenue du Mail** on the Place du Cirque (Plaine de Plainpalais). Lengthwise, the building slopes down the street. The auditorium is in the courtyard, set in the lowest level of the site (rue des Savoises). The auditorium floor is unusually placed: fully two stories below the entranceway. This is the predicament Saugey set out to solve with his design. He intention-

ally chose an arrangement which most would consider a difficult precondition and turned it into a catalyst for the building of an unparalleled cinema auditorium. The auditorium is perpendicular to the axis of the entrance and what it lacks in depth, it makes up for in breadth. Saugey turns a trip to the movies into an experience in promenade architecture (Le Corbusier, who coined the term, built the Clarté apartment building in Geneva in 1931, the year Saugey began his architectural career and a work which greatly influenced him). ——— A cavernous entrance brings the visitor into a long hall leading into the depths of the building. Right away, two ramps branch off, one leading up, the other down. Rounding a corner, both lead to yet more gangways, some practically horizontal, others rising or falling, which the visitor must step onto to reach the

platforms on which the two balconies are situated, one above the other. This creates, with the orchestra, a three-leveled room. Saugey's daring was to leave the cinema auditorium open and to trust both the definition of space and acoustic requirements to architectural elements other than walls. The ramps are covered in sound-absorbent wall-to-wall carpeting and act as a buffer between the auditorium and the entranceway. Their balustrades of wired glass shimmer in the soft light. The cinema's fascinating lighting sets off the boundaries between the different spaces. The general impression is one of facets, or perhaps even shards, which have been fused. The organic forms of the postwar period have been left behind and a crystalline world has been entered, anticipating the decade
to come. 84

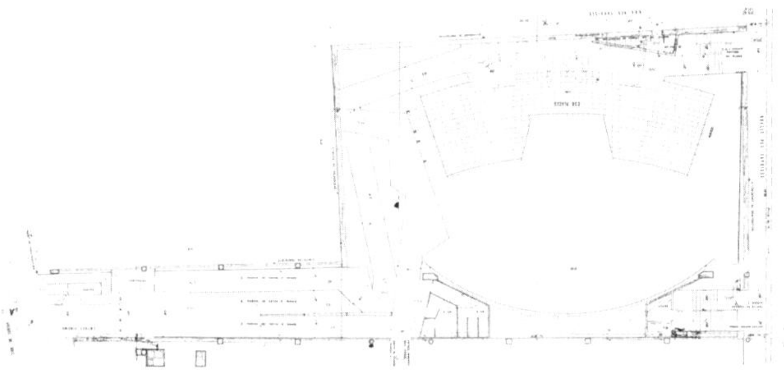

82 Movie theater Le Paris, lobby with ticket booth

83 Circulation ramps and cinema auditorium

Floor plan, cross section, and longitudinal section

photographs of the model

Balcony from above

86 Cinema auditorium, view from the ramp

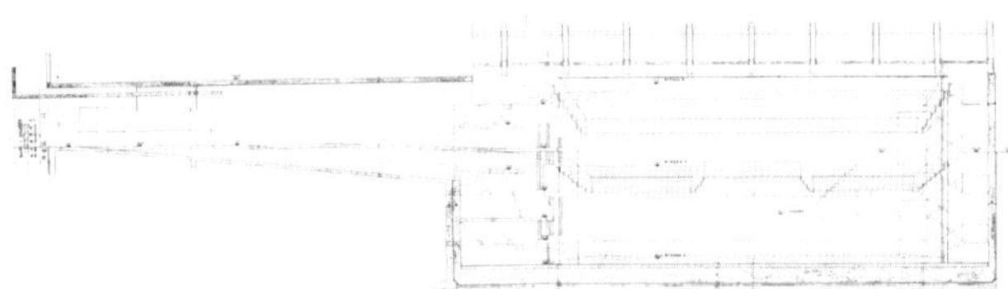

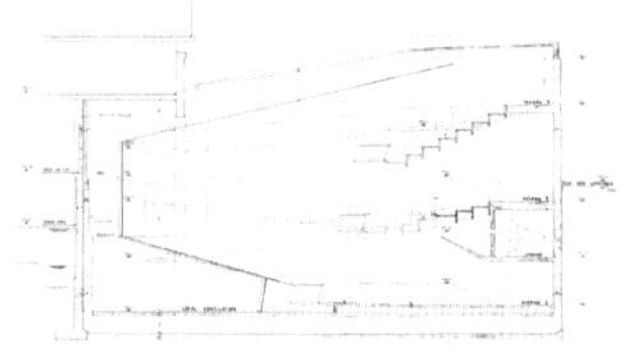

SORTIE

At the time when Swiss streets were still filled with numerous American cars, choosing to drive a French compact car for reasons beyond economics was making a real statement. Drivers of the CV2 waved to each other upon passing one another on the road as if to say “aw, let’em laugh at us!” In his **Citroën Poster**, **Karl Gerstner** emphasized not only the utility aspect of this car – its economy – but also its prize-worthiness. In **1958** his poster designed to look like a price tag combined photography and typography to appeal to his audience to be sensible, to make themselves elite in terms other than financial ones. ——— Gerstner’s **1961 logo** created for the **Bech company**, producers of high-quality entertainment electronics, also speaks to an elite group, but this time those with real buying power. At that time, consumers who were no longer satisfied with their “gramophones”, listening to their records through cables connected to their radio speakers, caught the Hi-Fi (high fidelity) bug, and were prepared to sacrifice space in their living rooms to set up a loudspeaker the size of a small washing machine for the sake of higher-level physics (Stereophonics were launched shortly thereafter). The white background of his poster, symbolizing the purity of the sound equipment, caught viewers’ attention in the hubbub of the billboard jungle. As related graphic idioms, Gerstner designed envelopes and letterhead, a record album cover and all of the newspaper and magazine advertisements for Bech.

Karl Gerstner, automobile advertising poster, Zurich, 1958

Karl Gerstner and Markus Kutter, poster “Nationalzeitung Basel”, 1960

BECH
ELECTRONIC
CENTRE

Karl Gerstner, Poster, 1960

Self-taught graphic designer **Siegfried Odermatt** has been creating outstanding works since the 1950s. We present here two works from the years before his collaboration with Rosmarie Tissi, with whom he shared an atelier after 1958, serving a diverse group of clients. ——— The **flyer** created for the Zurich company **City-Druck** in **1952** not only exhibits superior graphic design, but is also an important document in terms of the history of media or printing technology. It merges the modern technique of photomechanic reproduction with lino printing, layering three photographs of lino tools and the word “linol”, whereby only the ‘n’ and the ‘o’ can be identified as letters at first glance. The three remaining letters build two positive spaces and one negative one. The composition is fascinating and full of tension. ——— In **1960**, Oldenmatt designed

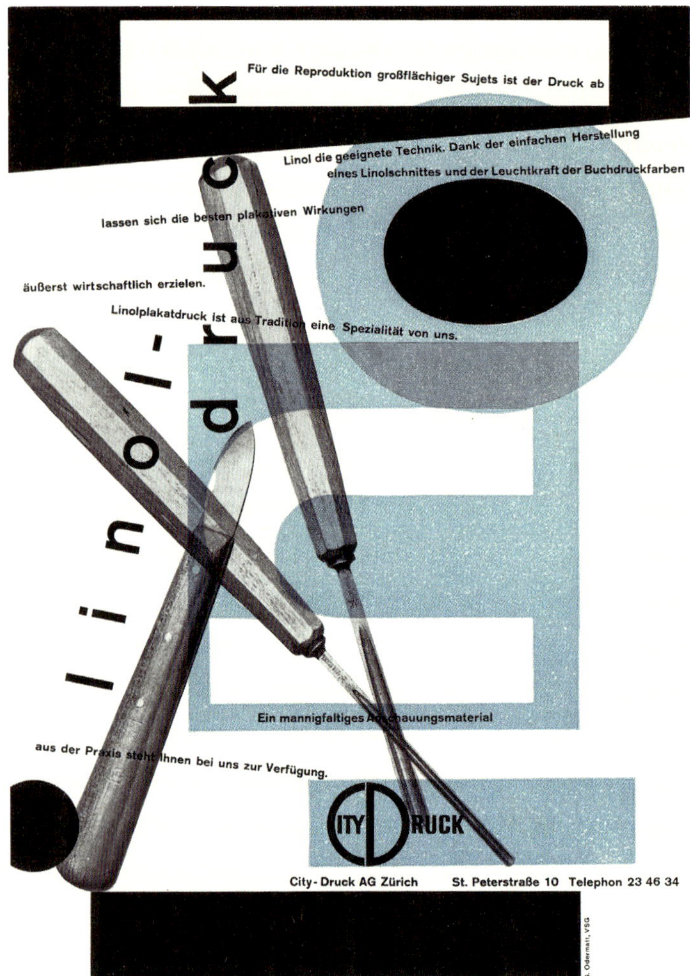

Advertising flyer, A4 format, Zurich, 1952

a series of slender **flyers** for the **Neuenburger Insurance Company**. They again reveal superior conceptual abilities in the joining of image and typography. On the automobile insurance flyer, a car has suffered considerable body damage crashing into the word “auto”. On the “water” damage flyer, the water’s refraction is used to visualize the drama of the situation and the necessity of preventative measures. The flyer for the event of a “broken bone”, being in French, makes us realize that one reason for the subtle typographic design of so many Swiss works of visual communication may lie in the fact that in Switzerland, graphic design must “work” in the country’s many native languages. Designers cognizant of this fact respond to this demand with a heightened understanding of the interaction of typography and photography.

Three advertising flyers for Neuenburger Versicherungen (insurance), 1960

Company logo on a delivery truck for office machinery, c. 1960

Nelly Rudin, poster for the “Saffa” exhibition, Zurich, 1958

Blotter for physicians, 1952

Before **Nelly Rudin** decided in the sixties to work as a fine artist, she was, as a young graphic designer in **1950**, an important freelance team member of the pharmaceutical company **J. R. Geigy's** design studio in Basel. She subsequently went on to freelance for Josef Müller rockmann's agency. She possessed a well-honed sense of form and color, developed in her studies under Armin Hofmann and Eidenbenz. Typographic, photographic, and graphic elements unite in her work to form an impressive whole. For J. R. Geigy, she designed numerous product packages, advertisements, and other printed matter such as, for example, a **blotter** that doubled as a **monthly calendar** for doctors writing prescriptions. After leaving Geigy, she designed a beer coaster for a brewery with a wordmark reminiscent of rising froth and a **logo** for an office equipment company in Zurich (shown here on a delivery truck). The official poster for the 1958 **SAFFA** exhibit (Swiss exhibit on women's work) is also her work. She used a photo by Jakob Tuggener for the invitation to a ball in Hotel Dolder. Nelly Rudin's importance after 1964 as a constructivist artist can be more than divined from her work as a graphic designer.

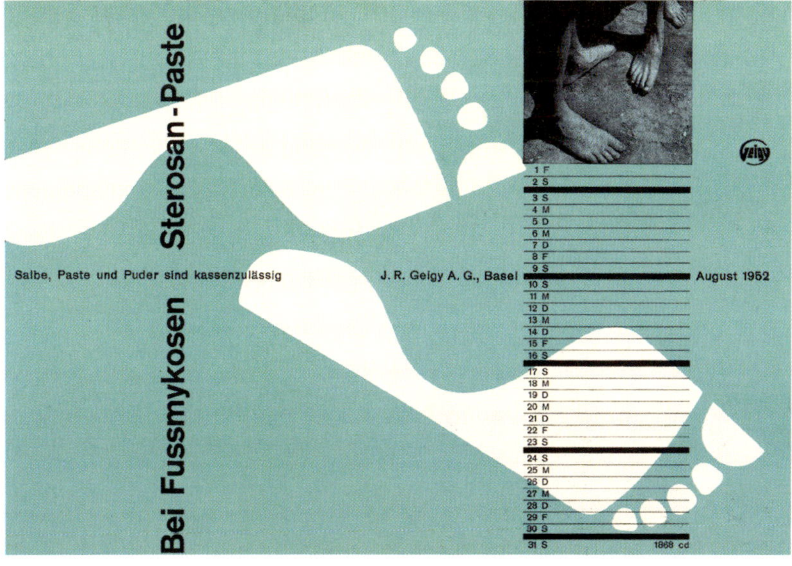

Zurich's stage for cabaret and pantomime, the **Theater am Hechtplatz**, was converted from a former fire department storeroom in **1959**. The architect, **Ernst Gisel**, responded wisely to the limited space and found the means to create just the right atmosphere for this project. The theater seats 220 to 260 people, depending on the size of the stage. Due to the scarcity of space, the elongated theater auditorium (18 meters long, 7 meters wide) has no side aisle, each individual row of seats is directly accessible via bi-fold doors. Therefore, the foyer possesses the same slope as the auditorium floor. At the same time, the paving on the foyer floor is a continuance of the outside urban world. The theater auditorium is painted red, the ceiling gray, the wooden partition wall in which the folding

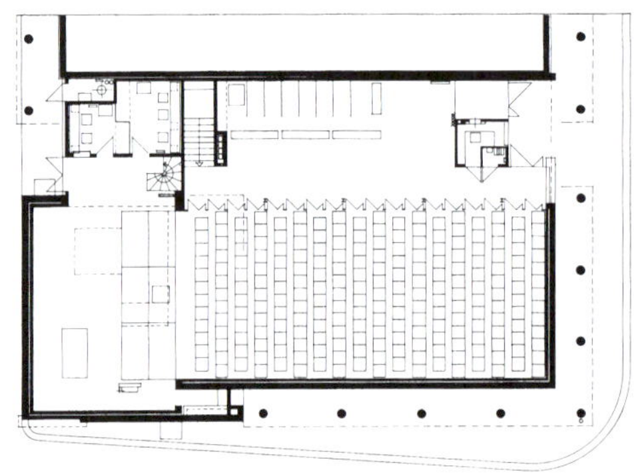

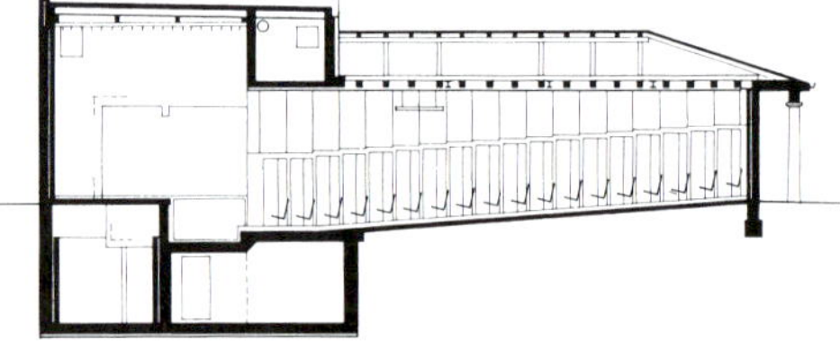

doors are set is black on both sides, the exposed steel supporting posts in the foyer are red, the darkly stained auditorium floor visually extends onto the black pavement of the foyer. By definition, a foyer is a transit room – in this case, between the city world and the theater world. Here, this transition is most clearly articulated with the barest means. The longitudinal foyer wall, made of gray sand-lime brick with face joints and wallpapered with posters, is also an element of urban space. A naked light bulb set in a brass plate at the end of each row of seats makes up the auditorium lighting. Here, too, there is quite a bit of charm, but no great commotion. This small theater is not a plush-covered refuge from daily life, but a meeting-place for the citizens of the city.

Theater am Hechtplatz, floor plan and longitudinal section

Two stills from the documentary film by Marc Schwarz, 2006

View from the lobby into the theater auditorium, photograph, 1960

Ernst Gisel designed this **vacation home** for his six-person family in **Rigi-Kaltbad** in **1959** on a ten-by-ten meter square plot. This is unusual in the mountains, where the landscape tends to polarize between the mountain and the valley faction. However, the square itself is hardly noticeable, only truly becoming evident from an aerial view. The roof has two angular eaves on the valley side and two gable walls facing the mountain. The living room reflects this order. It is the center around which all else revolves. It lies below the pitched roof area, which opens up in two different directions and becomes two stories high on

Axonometric drawing of the vacation home

View from the northwest

Interior with stove, and angular gallery above

View from the east, aerial view of the roof

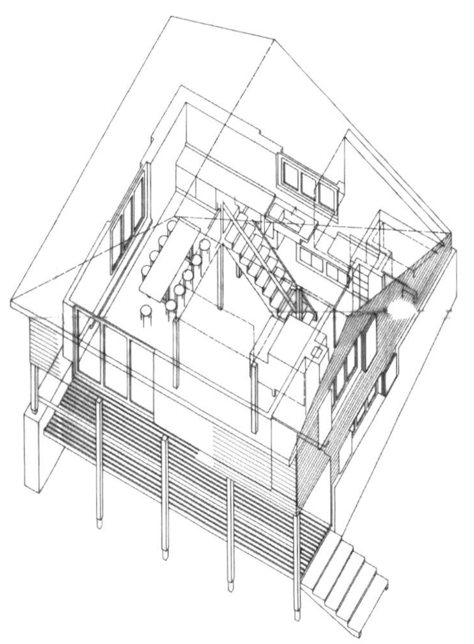

the valley side. ——— Overlooking the valley on the second floor are four small bedrooms and the master bedroom. They can be reached via a small angular gallery, which looks onto the living room below. An opening in the corner of the living room ceiling guides the warmth rising from the wood burning masonry stove to the bedrooms. The second most important room in the house is the wide covered porch on the south side through which one enters the house. It offers those who sit on it a view of the deep sinking of the Vierwaldstätter Lake or of the rising sea of fog.

Ernst Gisel took up the challenge of designing ateliers quite early (1948), quickly developing his own personal signature. But not until **1973** did he move his own office out of a former residential house built in the 19th century and located in the center of town into his new **atelier** on **Streulistrasse** in **Zurich**, a building which powerfully embodies the characteristics of atelier design. The three-story building is situated behind a remarkable residential building from 1960, also designed by Gisel, towards which the atelier turns the slope of its two (one behind the other) staggered roofs. The atelier provided working space for approximately thirty employees. The cross section of the building is sophisticated:

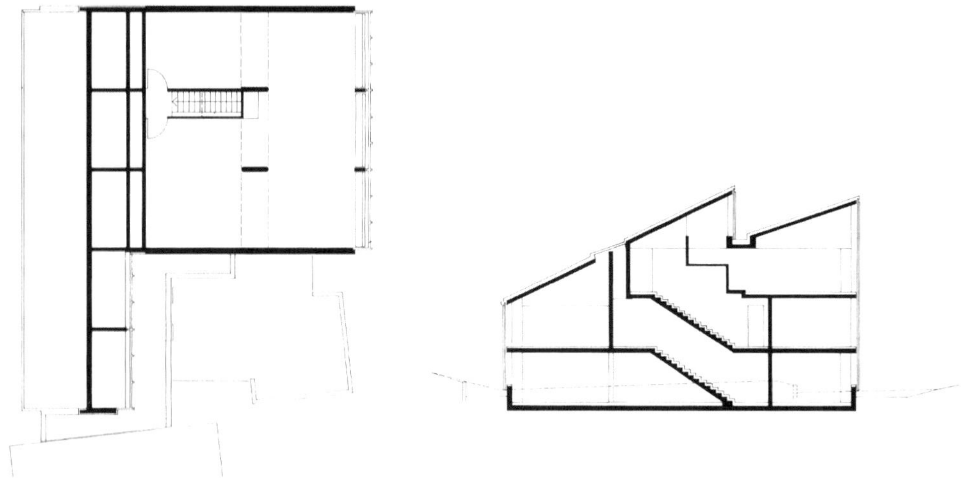

Not only do the northern glass panes allow daylight to enter the open plan office on the upper floor, but also the skylight in the slanted roof lets daylight into the ground floor. Concrete bulkheads support the building and give the rooms an alignment perpendicular to the axis of entry (Gisel brought the engineer Heinz Hossdorf in on this project, see also p. 32). Both gable fronts are roughcast and painted deep blue, the south and north elevations of the building are made of a white-painted metal construction. The north face consists of solid glass panes and tilting windows, whose characteristic side sheets serve both as rabbet and rain protector.

Floor plan of the upper floor and longitudinal section

Eastern elevation with entry courtyard

Office space on the upper level

View from the neighboring residential building

When **Fritz Haller** speaks about architecture, he speaks about building – about a process, not a form. His convictions diverge completely from what most people consider building to be. Haller speaks of one moment within the building process as truly fulfilling – when during assembly the crane lets the last girder down and from below you see and hear it snap into place in the grid structure which has already been built. Anticipation of scenarios, geometrically predetermined points in the plane and in space, the well-planned organization of space and time – these are some of the keywords. ——— With his typical consistency, Haller sees in this piece of furniture a small-scale component of an ever expanding periodic table of elements, leading from here to small buildings, to large buildings, to city planning and regional planning, and finally to the organization of the Earth's surface and, even further, to planning colonies in outer space. ——— Haller's vision is one of extreme clarity. It has even been called totalitarian. But this is unjust. Yes, it has totalitarian traits, but since they are structural models and not blueprints for daily life, it is not the same thing. Haller is interested in creating the preconditions for life, not shaping life itself. This is a fundamental difference. ——— "The

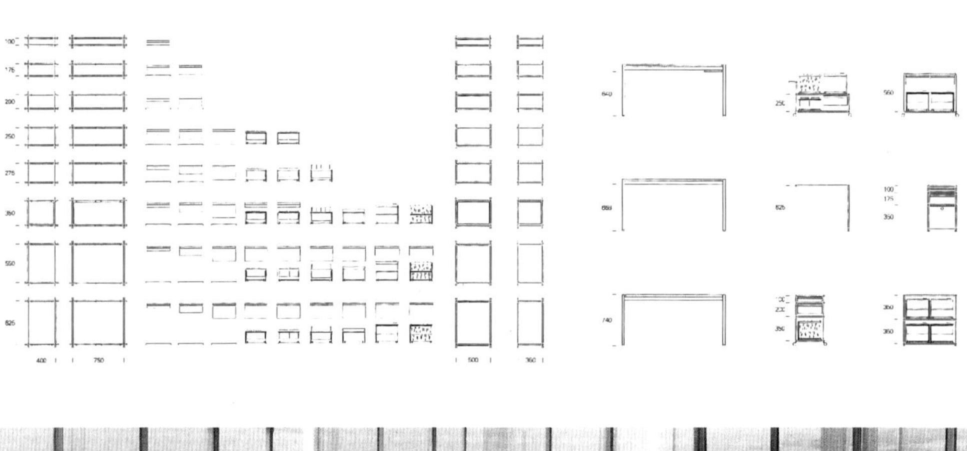

Modular furniture system USM Haller,
published in *Architectural Design*, London, 1967

The modular geometry of the USM Haller modular furniture system modular furniture system USM Haller is a complete kit and a closed system, consisting of the building blocks: supporting framework, paneling, fixtures, and fittings. A wide variety of objects can be built with the elements of the furniture building kit: filing cabinets or tool cabinets, open or closed, cabinet carcasses, "rollboys", reception desks, [...] grids for ceiling panels, etc. These objects can be dismantled and reconfigured at any time." (Fritz Haller 1988) ——— The centerpiece of the system is a chrome plated brass ball with three screw threads in the X, y, and Z axis of its Euclidian space. The balls are manifest points of modular order in a three-dimensional coordinate system. The actual framework is made of tubes of different lengths connected at the joints. Metal or glass panels – immovable or as doors – fill in the compartments of the lattice. A whole series of fittings turn it into a comprehensive system. ——— In fact, the USM Haller modular furniture system is a byproduct of Fritz Haller's work for Ulrich Schärer Münsingen (USM). Haller designed it in 1964 to furnish USM's new company building, which he had just built. A group of French visitors to the new building were so enthusiastic about it that USM went into production for a wider market.

Representation from the Fritz Haller exhibition *Bauen und Forschen*

Silberkugel
MÖVENPICK

The form-determining power of nature held a strong fascination for **Heinz Isler** from early in his life. Isler studied **civil engineering** at the Swiss Federal Institute of Technology in Zurich after the Second World War. One of his professors, Pierre Lardy, promoted a design method based not on mathematical calculation, but rather on **model tests** – that is, it focused on the form-giving effects of forces. Since his youth, Isler had had an affinity for nature and had been attracted to the visual arts. He thus occupied a field of tension between disciplines commonly held to be polar opposites. His early interest in naturally occurring shell forms (mollusks, snails, insects) led him to investigate forms that, despite (or perhaps all the more for) their resistance to mathematical analysis, demonstrated great structural integrity. He aimed to find forms that manifest the most pure compressive state possible. In the period between **1954** and **1963**, through his experimental curiosity, he worked out a morphologically based typology of forms that he classified as *shells (Buckelschalen), hanging forms (Hängeformen),* and

floating forms (Fliessformen). At an engineering conference in Madrid in 1959 Isler delivered a highly provocative lecture introducing his typology (as yet without the floating forms). It not only caused a sensation, but, due to its plausibility, had a long-lasting impact. ——— Isler's ***Buckelschalen***, or concrete ***shell structures***, have found the most widespread use. Quadratic in plan, they lend themselves easily to additive use as spatial modules. The concrete shells are very stable, despite an approximate thickness of just 8 centimeters. For industrial buildings, the broadly spanning perimeter support beams offer the additional advantage of their suitability as crane tracks. As a natural extension of his work on the shells, Isler also developed rooflight domes of polyester of up to 5 meters in diameter. In Switzerland and abroad, he built nearly 1,500 shell constructions. He praises their outstanding durability. Such shell structures are still being built today, although Isler himself acknowledges that the zenith of this type of construction is now past. ——— The designation ***Hängeform***, or ***hang-***

ing form, is typologically correct, yet misleading, as it most immediately brings to mind tension structures. Isler's so-called hanging forms are actually standing shells derived from the inverted shape of fabric or a similar membrane suspended from three, four or more points. Thus they are effectively pure tension structures converted to structures in pure compression. Examples include the highway gasoline stations at Deitingen-Süd (two such inverted hanging forms, each resting on three points) and the Sicli company administration building in Geneva (both built in 1969). The elegant structures Isler built to shelter tennis courts – 50 meters in length with a shell thickness of just 8 centimeters – also belong to this typology. ——— The transcribed section of a lecture delivered by Isler (p. 110) discusses his shells and hanging forms. In it the designer explains that the development of this new logic of form would have been fruitless without the parallel development of a method of construction employing high-quality, prefabricated, reusable structural elements.

104 Heinz Isler, Highway rest stop at Deitingen-Süd (Solothurn), just after completion, 1969

106 Top: Freezing experiment, c. 1955

Center left: Heinz Isler with the model of the pneumatic principle for shell structures, 1955
Center right: Experiment with structural mode

Bottom: Industrial building with shell structures (Blaser Swisslube, Hasle BE)

107 Sicli administrative building, Geneva, 1969, photograph of the construction site

Bottom: Framework for concrete shell to shelter tennis courts
Example of Isler's tennis courts, Marin

Heinz Isler on his work

109

HEINZ ISLER.
ON HIS WORK

How I came to Burgdorf? And to the subject of my life-work? An engineering colleague that I knew during my military service had an office in Burgdorf. He got the commission to build a concert hall in Langenthal, a cylindrical shell which was supposed to be rounded off in front here. He asked me about it. I looked at it, thought about it, and said I didn't like it. A cylinder is made of straight lines – but what is a straight line, anyway? There's no such thing. There is not a straight line in the entire universe. But there are curves – curves by the billion. A straight line is just a curve that can't decide whether its radius of curvature goes up or down. That is, a straight line is a curve in an unstable condition, and that isn't curve enough for me. I was against it from the beginning. I'm a man for curves – that's natural. So I thought, alright, I'm going to give this straight line a little bulge. I want a curve like this. I designed a curve for Langenthal, along the vertex, and then, crossing it, another curve and another. It was very easy to draw on the drafting table. We just had drawing boards, nothing else. I drew the plan really nicely with curve templates – my father was an engineer with the railroad and he had a whole set of curve templates, fifty different radii, wonderful French curves, you can do great things with them – and my idea was for every centimeter to have a different radius of curvature. I could draw well, so that wasn't a problem. For half of the shell I had to mark out coordinates. I chose 400 points that in the end determined the scaffold for the concrete work. Each member of the scaffold had a different form. They had to be specified to within three millimeters. So I had hundreds of points and when I was looking at this one point here and noticed that something wasn't right – a change in one single point – then the 399 other points also had to be changed. As you can imagine, that's an impossible task. You can't do it. I was at it day in and day out, for weeks, and at the construction site they were already building. One day the contractor's supervisor came to me and said, okay Isler, let's go, by six o'clock tomorrow morning I need at least three curves or you're off the job, and so on. So – onward with the curves. I had to give them three more and didn't even know whether they would fit together in the end. That's the way it went. It was excruciating! I'm someone who wants to think precisely and do his work well, and I was a wreck. I only knew one thing: That was the first and the last freely designed form. It just wasn't doable. ——— **BUCKELSCHALEN: SHELLS** Then, in the midst of my desperation, something happened, and that was a decisive moment. The story isn't made up, it really happened this way. I'm finally coming home at dawn – I had a garret in Burgdorf for a few months with a steel-frame bed in it – and I come home dead tired, I'm taking off my shoes and suddenly see something in front of me... the pillow on the bed had exactly the same form I had been searching for for months but hadn't found. It's an organic form, all curves, each line starting from a certain point and running from one curve into the next, and so on. Everywhere everything is round in all directions – organic – that's a natural form. What kind of organic form? It's a membrane, that is to say, a tension membrane; the pillow is puffed up with its feathers. At that moment I saw that that was the solution. What no engineer, what no person is capable of building, the chambermaid does every morning. She fluffs up the pillow and there it is – the finished form. And with my experience, with all the models I had built, I knew, okay, I'll build myself an artificial pillow, as exactly as humanly possible. I knew I had to work out the form down to a fraction of a millimeter, so that afterward I could enlarge it a hundred times, otherwise it would be wrong. That was the problem I was confronted with right at that moment. I don't know whether I stood there for one minute or twenty... Then I built a very accurate pneumatic model with a rubber membrane and a sturdy wood frame. Here, you see, I pump this up and all the originally straight lines become curves. Every curve is different within the overall symmetry, and now you see that it takes on the form of a hill, and that is the hill I was looking for. But there's one more thing that I'm free to choose: I can determine the height. If I stop here, it's this high, and if I add some more air, it rises a bit higher. And now, the result is a wonderful natural form that couldn't be more perfect or beautiful. This is a tension membrane, like a soap bubble is a tension membrane. The difference is just that the surface of the soap bubble has equal tension at every point and in every direction. Because it doesn't have any shear resistance, but always organizes itself such that the shear stress zeroes out. You can build a shell that way as well. After you increase the pressure and stiffen the shell with some material that you can pour over it and let harden, you can reduce the pressure. Then, instead of the pressure pushing upward, the weight bears downward. The sign has simply changed and I know: What was a pure tension membrane before has now changed into a compression membrane. I have one form for a square, others for different rectangles, another for a trapezoid, for a triangle, for a heptagon, for a circle, for an ellipse – every time I do this experiment, I do it under the same ideal conditions. It is also the smallest area that can be used to solve the problem. That means it requires the least material, has the least weight, the lowest cost and so forth: a whole string of good properties. And that's the case for each form that I experiment with. A whole world has opened up. ______ **HANGING FORMS** And then one morning I'm stumbling across a construction site, for a shell structure down in Langenthal, of course, that's how it always is... and now imagine those are the reinforcing nets that they use. I'm walking directly into the sun, it's rained the night before, everything is still slightly wet with the dew, and suddenly I stop dead in my tracks again. Look what I see right in front of me. You probably get the idea. Right there in front of me is exactly the same form that I had been looking for. Inside the square there, you see. And it was lit from behind and shining there so beautifully, a perfect form. The amazing thing was that it was just a hanging cloth, common sackcloth, that had been thrown there without a thought, and that gravity had pulled into this form. By itself, it made this beautiful form. It was clear to me now that this was far superior. Here, as flexible fabric, it's a pure tension form, and when I stiffen it and turn it over, it becomes pure compression. It was a cold winter, and I took fabric and hung it up and sprayed it with a fine spray of water. Very slowly we sprayed cold water on feather-light gardener's gauze. And look: here there's simply a cloth, a square cloth, and now when I turn it over, it creates the most perfect form, unbelievably beautiful, similar to this model. That's how we did it, a cloth sewn together, tied to four posts and then we sprayed it from above very finely and evenly with water. When I thought it was thick enough, we attached a cross out of roof battens to it, turned it over... a huge thing: three meters high. We put it back on the four posts. It held and it was wonderful, just lovely. I wanted to see how thick the ice was. Is that four millimeters or so, or how thick? And it was a just half a millimeter! It was covered in only half a millimeter of ice, and it held. And it didn't just hold itself, but an army lamp, too, that was even turned on, which I was able to freeze onto the fabric in the middle very quickly with a small piece of cloth. Simple as that! Even this concentrated load, and that was the most amazing thing about the shell! That was great. And that gave us this second possibility: with support at the edges and resting on points. Again: the area was double the conventional area. Up to then we had always used full edges, and now we could use edges or point-bearing. One point, three points, five points, seven points, unequal heights, unequally loaded, and so on – once again a whole world bigger than that I'd seen before. After that I made further observations... There are always new things you encounter along the way, things that you never would have thought of, and weren't searching for. You're accustomed to a certain order of things, and that's also a problem. Then you feel it and see it, then you can work with it. Since then we have worked in this area practically without interruption, solving this kind of problem, performing similar tasks. ——— **LABORATORY – FACTORY – CONSTRUCTION SITE** What I've shown you are rough models. But when you actually build, you need exact models. There are dozens of them in the other room. They have to be accurate to within a fraction of a millimeter at all points. Glue-laminated beams are made in the factory. Of course, cheap ones would be unusable; we use the most expensive ones, because they have to be the best. The best with these glulam beams means a roughly 32 meter span for this swimming pool in Thun. We make these curves 32 meters long, cut in the middle, of course. There isn't a straight line in them, only curves. Their curvature changes over their length and each one is different than the others because of the opposing curve in the other direction. They're all sections of organic forms. And I have to transfer them to the full-scale layout very precisely. Nails are hammered into a 30 by 30 meter area, from which the master builder takes off the shape of the beam. You make a pattern and that goes to the factory, where the boards that make up the beams are bent into the right curvature with presses and glued together. When they've hardened, you take them out of the presses and to the building site. But these things are expensive. A few years ago it cost 100 francs per square meter of floor area to make them. If I only used the beam once, it would burden the building loan at 100 francs per square meter. As you know, a roof of this kind costs between 200 and 400 francs per square meter of floor space covered. The scaffold makes up a significant portion of the cost. But if I make the forms good enough so that I can use them not just once, but ten or twenty times, the 100 francs is money well-invested. If I can use them that many times, it's incredibly cheap. I couldn't afford to do it otherwise. That's point number one. After that, roofing boards go on. They're flexible. You can span them from one beam to the next, spaced at half the width of the wood panels, that is, 25 centimeters, and then you have the raw form. Then the panels go on top of these. They're just laid down and lightly fixed in place. They take on a slight curvature. After the reinforcing steel comes the concrete, which is poured and carefully spread. ——— **NONSTOP** I demand a single pour, without interruption. You have to have that, or else the shell would end up like a bell with a crack in it that ruins the ring. And if you want them to ring – and these shells are like bells – they have to be made in one pour. We have very large shells, the tennis courts, for example, with 3,500 square meters of vaulted floor area, and you can't pour the concrete for a building of that scale in one day. But, of course, you can work day and night, a continuous pour. Or with retarding agents. But artificially retarding the setting of the concrete doesn't do it. When we have a concrete pour on a large project, we take everything available – weather forecasts for Switzerland, the region, Europe. Years ago the best thing was the air traffic weather service. They have to know if they can fly their little machines through an area, so they put out accurate forecasts. We registered weeks in advance, we received charts every day, hung them up one on top of the other, and you see, using these maps you can make your own weather forecast better than anybody else can do it for you. You just have to mark the different zones with colors and then watch. How is the front advancing each day? Then you can say the following day it will move this far. But that's not quite right, maybe it was 30 kilometers, then suddenly 35, or only 25 the next day. Thinking linearly isn't good enough. You have to take these variations into account, too. A week ago it was 40 kilometers per night, then 38, then 35, then 32, then 25 and now you can work out how far the system will advance in the coming day – okay, tomorrow it'll be 22 kilometers. So you know to expect a weather disturbance at a certain time in the morning, and another one at a certain time that evening. That's fairly precise. Your own weather report. You have the weather report, you have the raw information, but you put it together yourself. ——— **TRUTH IN POETRY** Then comes the decisive part. When the concrete has been vibrated, the laborers come to smooth the wet surface. They work on their knees and smooth the concrete by hand, using a common float. And that's what makes the difference; that's how you get virtually pore-free concrete in the top centimeter of the shell. Pore-free concrete, another miracle of nature. That's something very different than, for example, this standard raw concrete wall here, which is full of pores. This wall absorbs all the damaging stuff that we breathe out, and that's in dirt, in rain and so on, and these things cause the concrete to deteriorate – the acid and everything, the CO_2. But when we reverse it with the shell, which is only 8 centimeters thick, and the outer three, four, five centimeters are practically capillary-free, highly dense, with bubbles only on the inside – then the capillary action works in the other direction, not from the outside in, but from the inside out. That way, nothing gets in, neither liquid nor gas. So, oddly enough, we're the only ones who don't experience corrosion problems with the concrete, no carbonation, which can destroy whole facades and things like that. For this reason alone, all because of this work process. The process is very expensive; a square meter costs a sizable amount, I don't know exactly how much. But then we don't have any problems with it for five, ten, fifty years. We have a lot of old buildings at this point, all problem-free. Only the very fine cement skin on the surface gets washed away by the rain. All the shells are gleaming white at the beginning, wonderful. Then the rain comes and dissolves the lime, which disappears, exposing the aggregate – the gravel and sand. And we get washed concrete. You can see that well here, washed concrete on the surface, with the dirt from outdoors, and it's stable. We're down to the grains of gravel that last a thousand years without damage. That's something we didn't predict, we weren't trying for, but just happened that way, because we had to blanch the concrete to give it a nice surface and to seal it. It's a gift. I always say: When we work with nature, desig almost becomes a natural process. I follow natural laws as closely as I can. When I build the shell with correctly made concrete, it's also a natural product. And then the intellect comes inagain. I just have to prevent the structure from taking on tension, and that's accomplished with the right form. If I've prevented the structure from becoming tensioned, then I've prevented cracks. And where it isn't cracked, all concrete is extremely dense. So, by preventing cracks, I've prevented 90 percent of all problems that could affect the concrete. I haven't solved them – they simply never occur.

Heinz Isler, from a lecture to a group of German students, Burgdorf, Switzerland, June 2003. Transcribed and edited by Claude Lichtenstein

If there were an official ranking of shelves according to an index for the ratio of its own weight to maximum load, the **Lehni shelf** would be a strong contender for first place. ——— This shelf is not modular furniture, but rather a technically advanced interpretation of the well-known shelf type: load-bearing side walls with shelves set in between. It comes in two widths (44 and 88 cm), in four heights (53, 95, 180, and 220 cm), and in three depths (25, 33, and 41 cm). The shelves simply screw to the side panels, which feature rows of holes to permit finely graduated vertical adjustment of the shelves. ——— The shelf was designed in the spring of **1964** with a view to the Swiss National Exhibition in Lausanne (Expo 64); Max Bill, head architect of the section "Building and Design", commissioned the then 27-year-old designer **Andreas Christen**. His design has been in continual production by the hardware factory Rudolf Lehni ever since. The sheet-metal shelves are made of 1.5 mm thick **aluminum**, plain or powder-coated in blue, white, and black. Thanks to their being bent twice in the front and back, the side walls are stiff and sturdy, as are the shelves because they are bent downwards by 2.5 cm in front and upwards by the same amount in the back; a crimp at the right and left of each shelf creates the flange for the screws. Nickel-plated crosses at the back stabilize the shelf diagonally. Integrated bookends, slide bars for sliding doors and other accessories were developed later, but the nucleus of the Lehni shelf remains the conventional "rack", implemented in metal.

The Mühlehalde **terrace housing** in **Umiken** (near Brugg in Aargau) can be considered as a utopia come to life, from a time when the belief in radically new concepts played an important role. The name of the architecture firm that initiated the project, **Team 2000**, speaks for itself (architects: **Hans Ulrich Scherer**, **W.G. Strickler**). In **1964**, the idea of terrace housing as an alternative to the customary single-family housing developments was innovative and exciting. Scherer insisted on the design in the face of lethargic building regulations and property laws, neither of which had, at the time, foreseen a project of this kind. For the site of the prototype housing, the architects chose a steep site at the southern foot of the Jura Mountains. ——— The terrace housing is comprised of 16 levels clinging compactly one above the other on the steep hillside. The spacious single story units were financed by a combination of public and private ownership. The ceiling of the living and sleeping areas forms the terrace of the neighboring unit. Due to an inclination of 60 %, access was a central question. A funicular (an un-

Shortly after completion, 1964

2005

usual cable railway elevator) runs under a central horizontal walkway, stopping twice between the entrance to the housing at the bottom of the hill and the top level high above the Aare. Ninety meters of track cover a difference in altitude of 42 meters. The project has a linear structure, the funicular providing the backbone. From the lobby of the elevator station, one can reach the living units by foot. In the first stage of the project, a walkway leads to one apartment per level on the west side of the elevator. In the second stage, one more unit per level was added on the east side. The design of the terrace housing is quite remarkable, particularly in the first stage. The powerful form embodied by the multiplication of structural shells on the steep hillside is unlikely to be duplicated quite so persuasively ever again. Architect Hans Ulrich Scherer took on too much when committing to this adventure and departed this life at a young age due to financial worries. Nevertheless, shortly after his pioneering steps were taken, the architecture firm Metron was able to realize the planned expansion.

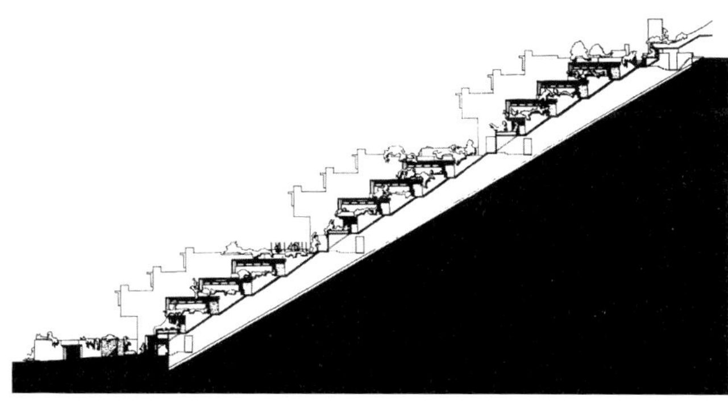

At night

Longitudinal section

The so-called School of Solothurn (Fritz Haller, Franz Fueg, Barth & Zaugg, Max Schlup) began around 1960 to realize a series of buildings, mostly at the southern foot of the Jura Mountains, which can be seen not only as the Swiss reaction to Mies van der Rohe and his Chicago school, but also to the postulate of prefabrication. Alongside industrial buildings, schools, and occasional residential buildings, the **Roman Catholic Pius Church** in **Meggen** (near Lucerne), realized in **1966**, is an astonishing and unique work by the architect **Franz Füeg**, also remarkable due to its location in conservative central Switzerland. ——— The supporting structure is almost that of an industrial building – a framework of unclad steel girders and an open web ceiling. Panels of marble approximately four cm thick are screwed to the framework, creating the church walls. 118

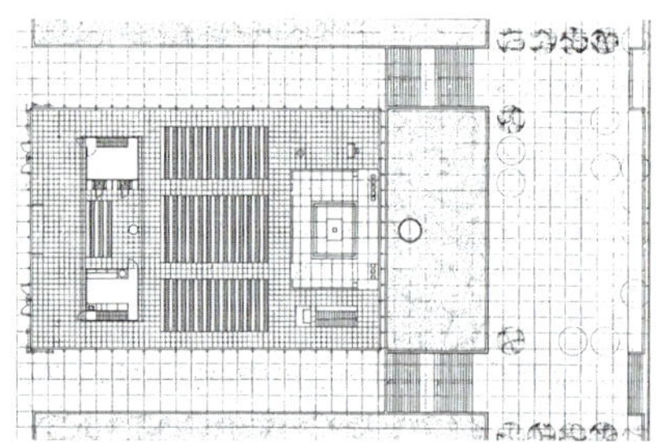

What appears in the daylight on the outside as an almost homogenous white surface is transformed into something completely different the moment one steps inside. Not only the direct rays of the sun, but also muted daylight easily penetrates the panels, revealing the play of the marble's veins, the internal color gradient, and the willfulness of nature. At night, the church even glows from within. The walls are vertical, the ceiling flat, the bell tower resembles scaffolding such as could be found in an industrial facility, the pews are anchored to the floor, even the organ loft is a testimony to rationalism. The design is based upon the tension between the steel skeleton and the timeless character of the veined marble (one could also say between the crystalline and the sedimentary) in back-lit or reflected-light conditions. From this it derives its impact.

Progress in inorganic chemistry provided new materials. Furniture designers responded rapidly and geared themselves towards the new possibilities. In Switzerland, **Ueli and Susi Berger** were at the forefront and designed the **soft chair** for the **Victoria-Möbel** furniture company in Baar in **1967**. The elasticity of foam permitted designers to give up the traditional separation of the supporting frame and the seat. The elasticity of foam rubber, correctly adjusted and tightly enveloped in a shell made of high-gloss plastic, resulted in a new fauteuil whose entire bottom surface rests on the floor. The curved concave-convex chair and an accompanying footrest (ottoman) are voluminous – a different kind of elegance than a decade earlier in the heyday of "good form"; now pop elegance in the pure colors yellow, red, green, and white. The distinctive grommets at the back (so that air can escape from the pores of the foam rubber) are sassy. This object was designed almost at the same time as corresponding pieces by Colombo. In contrast to Italy, Switzerland was not a design superpower, but it, too, had its cutting-edge talents. 120

A roll of packing material approximately 70 cm in diameter **(corrugated cardboard)** is cut into two equal parts with an S-shaped cut using a band saw. You now have two rolls, each with a flat end and a curved end. Each will become an easy chair – but isn't one yet. To become one, to be comfortable, the curved surface must also be hollowed out in the crosswise direction. Then, the middle of the curved surface is pressed down to create the bulge at the front of the seat. Now the seat is completed, but the previously flat bottom surface is domed. This dome is cut off with a band saw. The last step is to fix the edge of the cardboard along the side with glue, and the comfortable fauteuil is finished. This is the light-footed contribution of **Alois Rasser**, then a design student, to the legendary **1969 exhibition** on **"Paper"** in the **Museum of Arts and Crafts (Kunstgewerbemuseum)** in **Zurich**. About 50 such easy chairs were produced and sold cheaply during the exhibition. With a bit of luck, they can be found in well-furnished living rooms to this day.

In **1969**, architect **W. E. Christen** designed **stables** in **St. Luzisteig** (Graubünden) for the then still existing Swiss Army cavalry. They consist of two halls, one with six axes, the other one with four, at a right angle to each other. The construction of each individual hall axis is the essence of this design. Four columns of poured concrete support a long-span roof piece made of steel trusses that is 10.5 m wide and 21 m long. The main girders cantilever 5 m over the open court, providing a rain-protected area in the out-

Axonometric drawing of the roof structure of hollow structural sections

The succession of hall units with the angled lighting elements

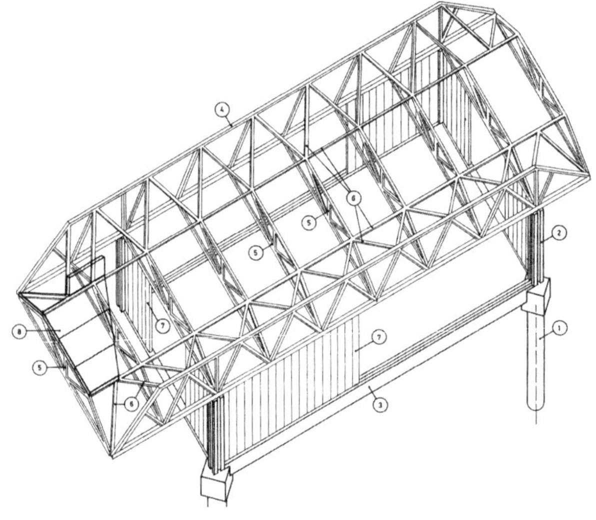

door space. Both longitudinal and transverse girders are plane trusses. Their connection at the sides called for space trusses. A band of transparent plastic panels on both sides of each hall segment brings daylight into the hall below; it covers the area where longitudinal and transverse girders form the space truss. The design was based on a highly flexible use of the space, which paid off after the army disposed of its cavalry in the 1970s. Today, the army still uses the buildings for other purposes.

View from the courtyard

Architect and designer **Klaus Vogt** designed a line of closets in **1964** named **"Squadra"** to be cut with a jigsaw. The line consists of three slender models for use as a cupboard or wardrobe and two wider models which can be used as a bar or a secretary. All models begin with the front panel of the closet, a 12 millimeter sheet of plywood, out of which the doors, drop-leaves, and drawer fronts are cut. The remaining area of the front panel is then mounted onto the body of the closet, acting as a brace so that the side panels need only be 10 millimeters thick. Depressions are molded into the front so that the doors can be opened. The hinges play an important role in the final statement of form. All models have a height of 140 centimeters and, depending on the model, two widths (60 cm and 86 cm) and two depths (43 and 60 cm). The design allowed for the additional use of the cupboard as a packing case, it has handles for this purpose. Since the closet has no foot, the user could choose from a left or right opening door in all three slender models, depending on which end was to be the base. ——— A trained boat builder, Vogt has a strong interest in the correlation of construction and form; he unites the areas of furniture design, interior design, and architecture. Squadra also testifies to the influence of Le Corbusier on much of the design of the sixties. Its dimensions can be found in Le Corbusier's logarithmic scale of proportions, "Modulor". The design of the cutaways on the different models are also reminiscent of Le Corbusier's early Purist paintings. ——— The design is a still youthful witness of the Pop Art era, the varying models for sale again today at the high-end furniture store Marghitola in Lucerne. The front panels, however, are now cut with a water jet cutter, which has superseded the jigsaw.

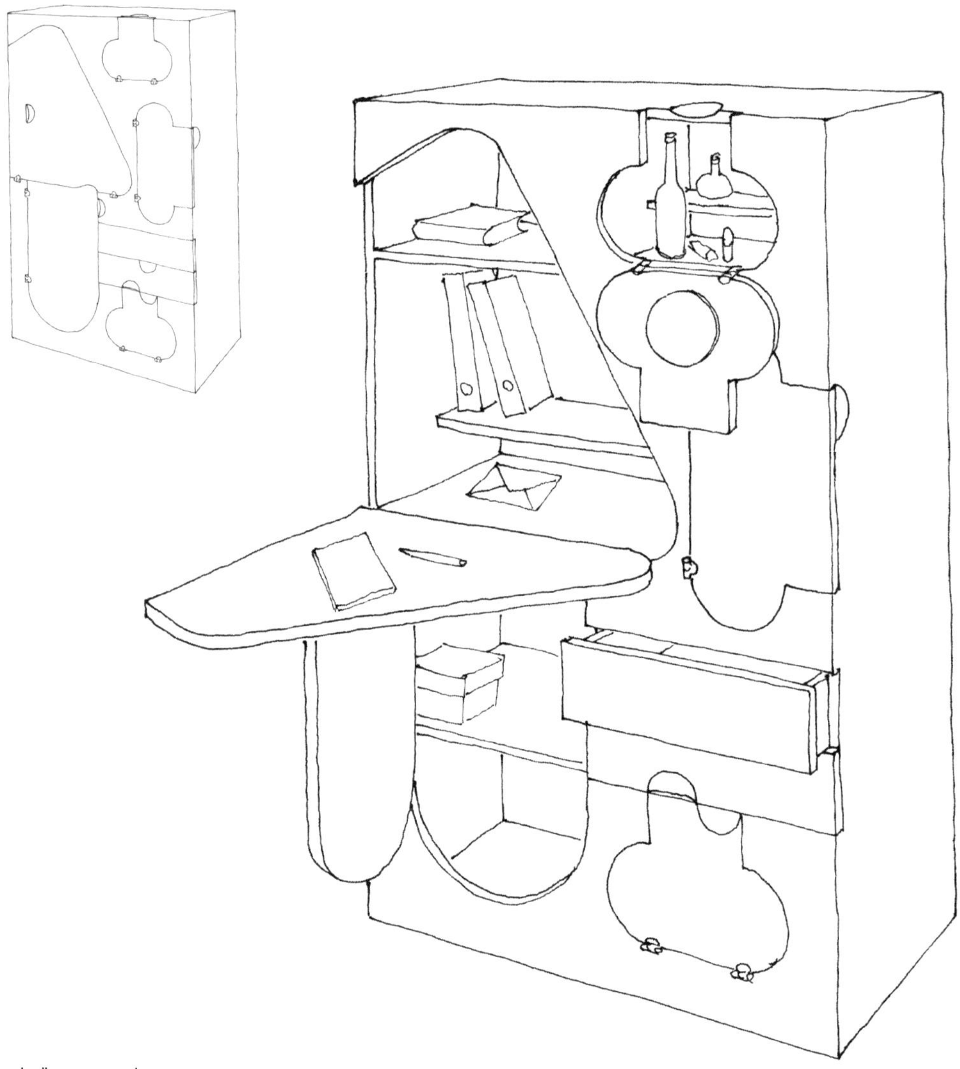

"Squadra" as a secretary

1974–1989
REFLECTIONS AND ALTERNATIVES

It would be interesting to ascertain when an era gives itself a name. Most such designations are, of course, applied after the fact and by others (Baroque, Classical, Avant-garde, etc.). The era in question here, in contrast, does have a name for itself: *postmodernity*. Here, the demarcation which each designation is supposed to achieve vis-à-vis the others occurs even within the word itself. But was that which preceded it modernity? No, but the 1960s and early 1970s failed to call themselves anything in particular, perhaps because everyone was much too busy to ponder what they actually wanted to achieve. Is that so terrible? Didn't Sigfried Giedion describe the hope to "develop a modern tradition" around 1940? The 1960s, especially the years around 1970, brought us this tradition, if one sees it as the practice of using and transmitting what one has learned – an era that does not question where it came from nor where it is going. Unfortunately, the result was not as good as what Giedion had expected. The years around 1970 are surprisingly fruitless – no comparison with the state of affairs a decade earlier. The **architecture** is often coarse, overly professional, and skillful in an uninspired way. It uses robust methods more effective for achieving quantity than quality. The same holds for **design**; true ideas are rare, and many furniture designs are remakes of earlier ones, only employing thicker boards or more massive steel tubing. Most **graphic design**, too, is routine, undifferentiated, and mass-produced. Books and magazines from this era speak for themselves.

For the period preceding 1975, we selected examples that stand out from this busy mediocrity. Just as in other sections of this book, we take the liberty of deviating from the precise dates of a work's creation in order to accentuate what makes it exemplary. Dating back significantly before 1973, both the house in Verscio and the swimming pool in Bellinzona pave the way, in the issues they tackle, for a new attentiveness during these intense years of rediscovery. This makes them all the more important; we "postdated" them here because they are exemplary for the interest in new themes.

Not least because the economy is faltering, the question of "where to?" arises emphatically, and with it the question "where from?" Where are we going if progress isn't progressing? And what kind of journey will it be if we are suddenly unsure of where we were actually trying to go? Even these questions themselves are findings. After the long years of overheated activity, there is once again time to think.

Unexpectedly, interest in history reawakens after a long hiatus, reaching back further than the 20th century. People are astonished to rediscover the Italian Renaissance (Palladio), Mannerism, and the Baroque. After abstract concepts of cybernetics (infrastructure etc.) were at the center of interest around 1968, the question of form pushes its way back onto the agenda. This stance is manifested with particular vigor in Ticino with its references to the Italian debate around Cattaneo, Grassi, and Rossi.

Considering that postmodernity is not a label from the outset, but points to a glaring gap in theory, the period's beginnings are impressive. All at once, it is acceptable to talk about form again, about the thickness of materials, effects of reflections, and nuances of color. Suddenly, the materials selected are more sophisticated: root grain veneer, decorative molding, and painted marble effects, mirrored walls and concealed lighting, rooms designed so that

the floor tiles fit exactly, aligned along a central axis. The quotation replaces the proclamation, and where something is proclamation, it expresses a doubt or contradicts *convention*.

But what is new and exciting in the first half of the 1970s loses its freshness fairly quickly and wilts into a somewhat sourish catechism. Viewing magazines from the 1980s reveals that for most of them, their "best before" date has long expired. The number of examples appropriate for this study is surprisingly low. The reason is first and foremost that these postmodernist years had a very pedagogical approach, and that real ingenuity seeks to liberate itself from so much well-meaning didactics. The difference between ingenuity and playfulness – mentioned briefly in the introduction – can be grasped here.

The swimming pool – **"Bagno pubblico"** – by **Bellinzona** is considered the key architectural opus in Ticino; it triggered the blossoming of architecture there and the international attention it received after 1970. That the architects **Aurelio Galfetti**, **Flora Ruchat**, and **Ivo Trümpy** won this competition in 1967 was a kind of signal for a "critical" architecture and against the pseudo-Italianità being built for the growing number of "tedeschi", builders from German-speaking Switzerland and Germany, from beyond the Alps. The "Bagno pubblico" exemplifies a new understanding of design and an awareness of architecture's interventionist role in an overused cultural landscape. ——— Bellinzona makes no attempt to interpret the swimming pool as a natural place (lawns, pools, and little hills, with a few changing pavilions and a kiosk in between), but rather sets it into the corrected Magadino plain as a built structure. An elevated walkway more than 300 m long and 5 m high, made of cast-in-situ concrete, connects the road with the embankment of the diked Ticino River. By raising the horizon, the walkway enables pedestrians to experience the plain as a cultural land-

Aerial photograph from the east

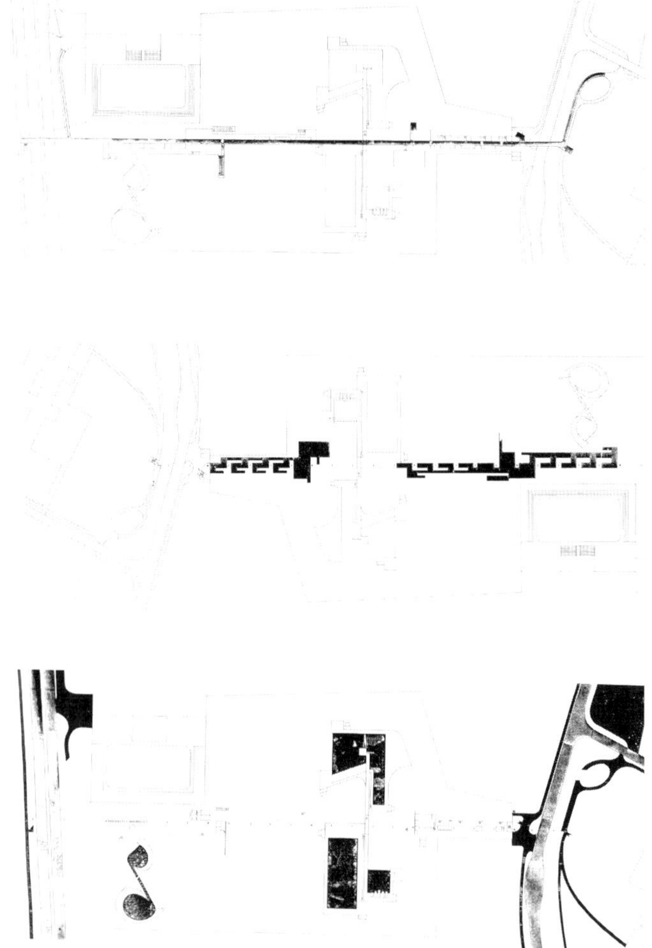

Floor plans (from the top):
Upper level (elevated walkway), total length 300 m.
Mezzanine (changing rooms), ground level (lawn and pools)

scape and the bordering mountains in contrast to it. Due to its large scale, the swimming pool with its "viaduct" also takes on a symbolic function in the landscape akin to that of the medieval Urn dams (for instance Castelgrande in Bellinzona). ——— The entrance to the swimming pool is located about halfway across the walkway, but one story lower, and is reached via a ramp. Here, at the first-story level, are the changing rooms and cloakrooms, on a deck made of wooden slats and with a bolted steel supporting structure. The women's changing rooms are behind blue partitions, the men's behind yellow ones, which punctuate the way like signals lighting up. On their way from everyday life to the pleasures of swimming, the bathers climb down another flight and arrive on the lawn and at the pools. ——— All the elements create positive tensions within themselves and with each other: the viaduct through its asymmetrical access ramps, the cloakroom decks through their arrangement on either side of the walkway, the pools through their differentiated form in dialogue with the datum of the "viaduct", and the diving tower by its diagonal thrust as it grows upwards into space.

For **Bruno Monguzzi**, the insights of the early 20th century have retained their validity. Quoting El Lissitzky, he considers himself a designer in the tradition of modernity: "Typographic design must express visually what an orator expresses by means of his voice and the language of his gestures." But how does one move the material to speak? How can one discern what the message is? That is true art. For the graphic designer does not depict anything visible, but makes something visible in the first place, in analogy to Paul Klee's famous sentence, "Art does not reproduce the visible, it makes visible." ——— So **visual communication** is also communicating the visual. To Monguzzi, designing a **book** or a **poster** means finding the best form for the topic at hand. Or, more precisely: finding the formal implementation that corresponds to the content completely. To listen to him speak about the design process is to hear a report on the activity of a detective (see page 140). ——— Monguzzi designed a series of publications on various neighborhoods on the outskirts of Milano consisting of two different kinds of texts: historic accounts of the city's history and urban development as well as sociological studies on life in these areas. The clients originally had two different books in mind. Instead, Monguzzi proposed a typographical concept that intertwined these two realities closely. The contributions by the specialists in urban architectural history were arranged in vertical columns and set in

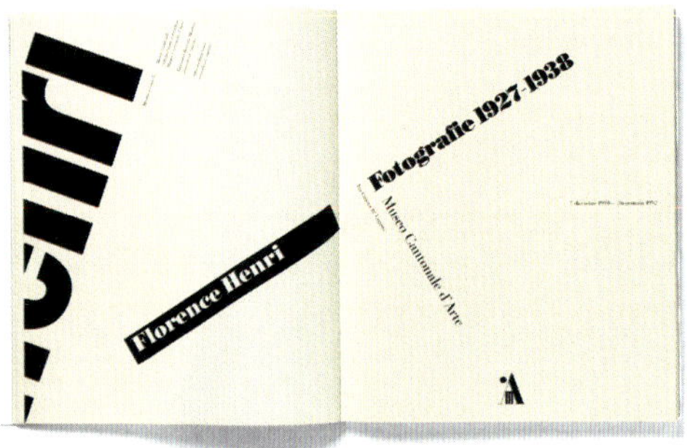

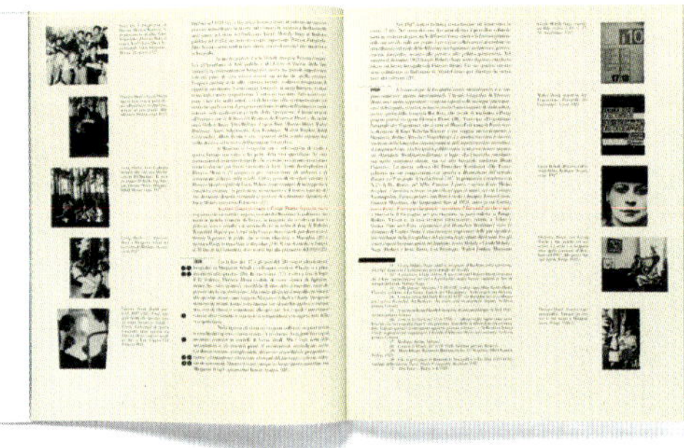

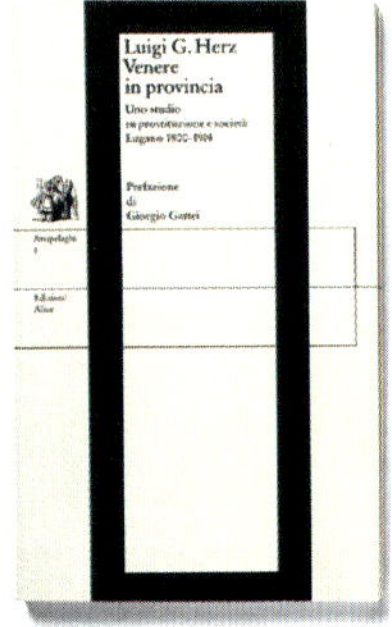

an Antiqua font, the interviews with people living in these areas and the sociological contributions were set in other typefaces and laid out in horizontal fields that covered the entire double page. The painstakingly constructed relations between the maps and the photographs on these spacious pages also show how "architecturally" Monguzzi arranges the material and how he uses the placement of different image configurations to cause the elements to interact. This is not only a question of aesthetically skillful layout – it requires a comprehensive understanding of the material. ——— Monguzzi considers himself something like an image engineer. Time and again, he took the photographs of his pictorial elements for posters or advertisements himself and constructed a whole out of the photographs and the other elements. He developed a comprehensive concept for several series of hardcover and paperback titles for the Ticino non-fiction publishing house Edizione Alice (The series deals with sophisticated topics in the fields of medicine, psychology, sociology, etc). Viewed in total, the various series – related, but differing from one another – project a strong identity of this scientific publisher to the public and a differentiation of the series to the inside: by means of the color of the paper, the arrangement of the bars or frames which structure the layout, the circular elements that turn into bold horizontal stripes on the spine. See also the statement by Bruno Monguzzi on page 140.

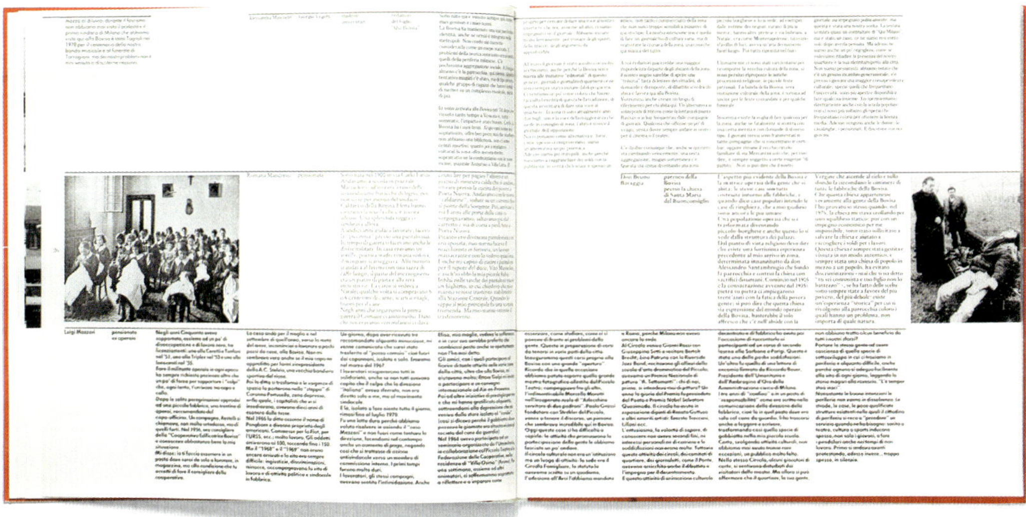

Catalogue *Florence Henri*, Lugano, 1991.
Title page and a selection of double page spreads

Cover designs for Edizione Alice (Viganello)

Publications *Milano Zona 5* and *Milano Zona 7*, 1982/1993. Double page spread

Florence Henri
Fotografie 1927-1938

Lucia Moholy
Ritratti al Bauhaus

henri

florence

Mercoledì–Domenica 10–17 Martedì 14–17
Lunedì 24, 25, 26, 31 dicembre e Capodanno chiuso

7 dicembre '91–26 gennaio '92

Museo Cantonale d'Arte Via Canova 10 Lugano

Bruno Monguzzi, exhibition poster "Florence Henri", Lugano, 1991

Bruno Monguzzi, exhibition poster "Anwesenheit bei Abwesenheit. Fotogramme und die Kunst im 20. Jahrhundert", Zurich, 1990

BRUNO MONGUZZI. VISUAL COMMUNICATION AS A QUESTION OF TRANSLATION

I knew that I wanted to use this photo, because there were two exhibitions in the Lugano Museo cantonale d'arte, the main exhibition on Florence Henri and a small exhibition on Lucia Moholy's Bauhaus era photos. And the portrait of Florence Henri is by Lucia Moholy. OK, so I had this photograph and I had the idea with the mirror – which I wanted to use because Florence Henri had done such pioneering work with mirrors. That's all I knew, and all I could know at that moment. And in cases like this, I don't pick up a pencil or go to the computer, because I don't have anything in mind to sketch. So I have to play with the mirrors. Anna, my wife, brought four small mirrors home from the store for me. So I took this portrait photo in the size I happened to have at hand, taped it onto a piece of cardboard, leaned it against a bottle of mineral water and started to play with the flare... then I experimented with two mirrors. – Nothing, nothing at all. I spent hours on it, but I didn't get anything I could use. I felt depression coming on and I thought maybe I'd never finish. Then I pulled myself together and I began to ask myself, what's the reason it wasn't working, why weren't the flares producing halfway fascinating or mysterious effects? Then I said to myself: "Is it maybe because I myself can see what I want to reflect in the mirror – maybe the right idea is to reflect something I can't see?" Then I xeroxed this photograph – I thought I'm not making the poster yet, I'm just trying something out – I attached the photocopy to the back of the cardboard and propped the mirror up behind it, and I began to see this and that, and the other; and I saw that it began to be magic. But at the same time it was still rather primitive, because I saw a picture in front and I saw the mirror that reflected the same picture. So then I took a second mirror and I set it to the side – and in this moment I was on the right path, because I saw that the second mirror reflected the reflection and that the portrait photo of Florence Henri was now in the right direction and no longer inverted and I realized that I needed two mirrors. – No, that's wrong, I needed three mirrors, because instead of the cardboard with the photograph on it I also wanted a mirror that I wrote the name of the exhibit on. Because from the very beginning I wanted to find a composition which would do without additional typography, I was looking for a solution that would have the name of the exhibit already in the photograph. Of course I knew that the cubists had already done stuff like this, and Florence Henri originally had a cubist background, which probably explains her fascination with mirrors, because mirrors deconstruct reality, and that's also what she does. And in that moment I had the idea of separating her first and last names from each other, putting the name Henri on the back of the mirror and her first name, Florence, somewhere else. The first and last names should come from different mirrors. It wasn't until this moment that I reached the level of playing with dimensions, the dimension "Florence" here and the dimension "Henri" there... I wrote these names on a transparency which I of course attached the wrong way around, so that it would be the right way around in the mirror. Just like this, with the bottle of mineral water, the tape, a powerful industrial lamp, etc., I photographed the composition as a test, and the powerful lamp created this fascinating reflection here at the front. I never would have thought that the mirror would cast such triangles on the table, just because of its thickness! Really, these lines, and the triangular shape here and the shadow of the mirror – that was all really there and nothing was added after the fact. Sure, later a professional photographer came and took a picture of the composition with a professional Sinar camera, and we had a large format negative for the poster production. But the whole thing was really done here at home and not in a photo studio. The whole process. I was searching for an idea and I wasn't sure if I was going to find it and when I did find it, the solution was a tangible model that only had to be photographed. ——— You move things, you pay as much attention as you can while you're doing it and then you can almost say that you were led by chance. You have an idea in your mind, but it's not an image that you have in your mind, it's an idea that you have to find the fitting image for. That's why you can't just sketch it. You can't anticipate it. You have to move things and pay attention to the phenomena that happen. And these phenomena lead you to a hypothesis or to a series of hypotheses that emerge from one another. It's like creating a chain of insights. That's why it's so important to really know your point of reference. ——— And what's most interesting in this process is: The designer radically transforms his own person in this process, in order to become another person, but in exactly this moment he becomes himself, because in exactly this moment he is a player. ——— I am convinced that the designer is a translator. A good translator loves literature and writing, but he writes the same sentence maybe fifty times until he's sure that he's now truly transferred this sentence by Shakespeare into Italian; it isn't Bruno Monguzzi in Italian, it's Shakespeare. For this he needs Monguzzi's brain, his sensibility, his intellect, his culture; it needs all of Monguzzi, so that it can truly become Shakespeare.

Bruno Monguzzi in a film by Marc Schwarz,
December 2005

(Continued from page 59). The **TEE trains** were in service until **1988**, when, with the introduction of the **Eurocity concept**, the exclusively first-class trains were abandoned. The Swiss Federal Railways (SBB) decided to have the five TEE trains redesigned and rebuilt. In 1987, Uli Huber, SBB's head architect, commissioned architect **Franz Romero** with this renovation. The design specifications were difficult to fulfill: The number of seats was to increase from previously 168 to 231, the dining car was to be abolished in favor of meal service for passengers at their seats, as was customary on airplanes. To compensate for the loss of the dining car, a 12-seat lounge was to remind travelers of the elation of trips on the TEE. Design, planning, and realization were to be carried out in little more than a year's time, and this feat was accomplished in the SBB's then still existing main workshops, where outstanding professionals of a variety of disciplines were assembled.

Romero's design for the exterior featured a dark ribbon painted at window level whose color contrasted with the remainder of the train

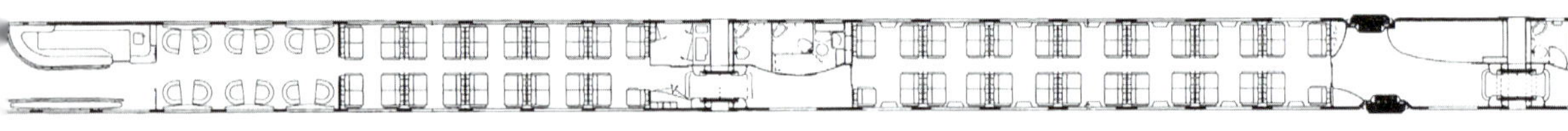

Two trains in the workshop
On the left: the TEE before rebuilding

Eurocity, seating arrangement with fold-out tables

and gave it a modern appearance. The interiors of the cars were fitted anew in their entirety; more than just the seats were to be replaced. The architect succeeded in permitting the culture of appealing visual and tactile detail to live on in the new post-TEE era by transposing the stylish and vivid form of the old seats into a new idiom, as well as by his choice of materials, for instance, by upholstering their sides with Stamoid. Designing all screw connections to be visible is one more example. An aluminum container for suitcases was built into the space between the seat backs. The halogen reading lamps with which the lateral luggage racks were newly fitted, were also successfully integrated in the existing structure. Because of an increasing number of technical problems, the five new trains were taken out of service around 1995, and all but one were scrapped. This remaining train is to be re-transformed into its original 1961 state at the behest of a rail enthusiasts' club. Unfortunately, Romero's remarkable achievement will be remembered by posterity only in pictorial form.

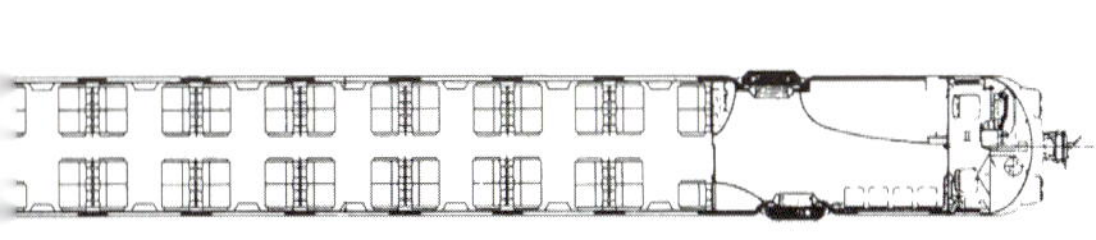

Lounge car with new fauteuils

Floor plan including lounge car

The **Snider House** in the little village of **Verscio** (Terre de Pedemonte, Ticino) was a clear statement by the architects **Luigi Snozzi** and **Livio Vacchini** against the false Ticino Domestic Revival with its "Klosterputz" and covered patios. Two old houses already stood on the plot. The group of buildings shows remarkable coherence. The new building differs from the existing ones in its materials, its constructive form and its stylistic idiom. By creating a two-story glazed opening in the facade to the interior yard, the architects achieved a concentrated exchange between old and new – concentrated precisely because this relationship is not constantly evident, but must be activated in one's own perception. Mutual respect for the qualities each possesses is the foundation for this exchange. The stringent terracing of the common yard by means of straight flagstone walkways and granite walls is and example of "convivial" thinking. The ensemble strives to make historical forces visible: from the pre-

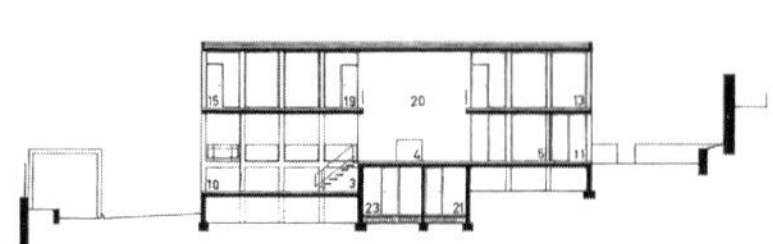

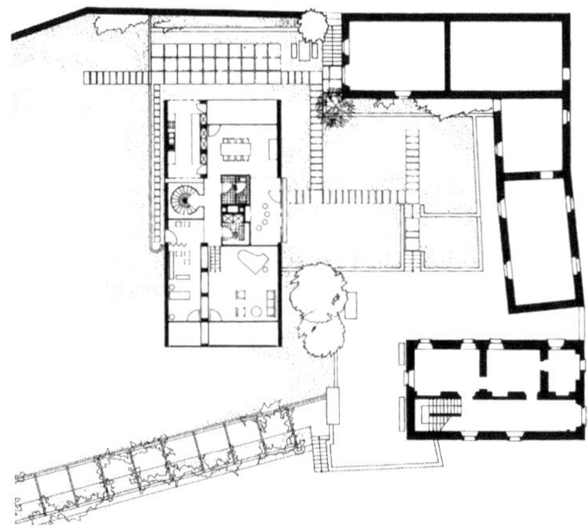

Snider house, floor plan with the old houses and the yard, longitudinal section

View across the common yard to the long side of the house

industrial period to the modern turning point of the Enlightenment, to the societal demands of modernity. The rational, flat-roofed house with smooth surfaces covered in light-colored stucco stands in the tradition of the “Neues Bauen” or “Architettura razionale”, as Alberto Sartoris phrased it in 1931. It is a particularly successful example of the ways in which rationality and emotional intensity can relate to and enrich each other, if one does not falsely assume an a priori contradiction between the two. The floor plan of the house comprises two oblong zones with a row of columns between them. This organization permits different functions to fill the individual spaces (areas that flow into each other in a structured manner, or closed-off rooms, the spiral staircase as a contrasting element) or to leave some of them open, so as to create exchange and visual connections between the two stories and across the yard.

South elevation

As the organizer of **"Jazz in Willisau"** (concerts and the annual late summer festival), for nearly forty years, **Niklaus Troxler** has earned a reputation in the world of jazz and improvisational music. "Knox" Troxler, as he is called in the community, based his lifework as a graphic designer on his acoustic passion, as well as lending it visual expression. His numerous works spanning this long period are illustrative and typographical in nature. ——— The **concert poster** for **McCoy Tyner** is a typographic masterpiece from the early years (1980). In a way that reflects the artists' shared genius, it advertises the pianist Tyner's powerful music, rearrangements of singable melodic lines as sound surfaces, structuring them into interlocking rhythmical blocks (one is inclined to say: from a point to a line to a

Festival stage Willisau, 2003 (showing Matthew Shipp)

surface to space). The poster is a two-tone silkscreen print in the complementary colors of yellow and violet; its letters are not outlined, their positive surfaces have the same color as the negative surfaces of their neighboring letters, and this very detail makes the image so intriguing: each character is distinctly legible. The size of the letters varies from line to line; the lines cover the entire width, leaving no interior boundary between a "figure" and the edge of the poster. The overall result does not have the effect of a "figure on background", rather that of an intended floating state of equilibrium. ——— The **concert poster** for **Marty Ehrlich (2006)** is a delicate work with a completely different feel both conceptually and in its
 execution, yet it also belongs to Troxler's range of expression.

LP cover: "Cecil Taylor", 1982

Niklaus Troxler, concert poster Willisau: “McCoy Tyner Sextet”, 1980

The New York Times

COLLEG… OPPOS… CALL TO UPG… ONLINE SYSTEMS

DAS… OF M…ICO TAKES …RASHING AS STORM STALLS

MANY ARE LEFT AT RESORTS

28. Jan. 06

Unseen Enemy Is at Its Fierc… In a Sunni C…

…k Case Renews Question… on War's R…ional…

Thous… s of Demolit…ns Near, New (… Braces f… New Pain

Niklaus Troxler, concert poster Willisau: “Marty Ehrlich Quartet”, 2006

In **1979**, architect **Ivano Gianola** built **Casa Bernasconi** in **Balerna** – a village in Mendrisiotto located at the most southern point of Ticino. Since then, "postmodern" might have become a bad word, but actually only because of the opportunists who misused it and saw an easy bait in ironical tricks of all kinds. This narrow house is of a different class. It subtly exhibits some of the themes of the architectural discussion of that time. The clear language of the house, its geometry, its "persona" as a small palace, the décor, and last but not least its inner spaces, cultivate a witty ensemble. The house

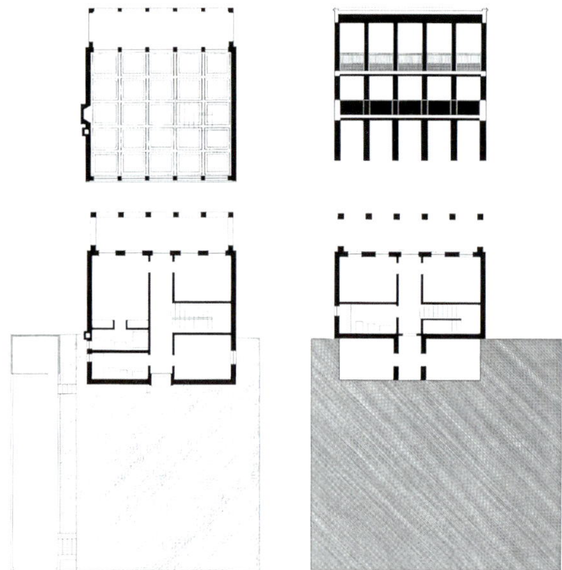

Floor plan upper level, entry level (left)
Garden elevation (upper right), garden level (lower right)

Studies for painting the facade, 1979

is not as small as it appears to be from the street. The main side is three stories high and has a beautiful loggia. From the street, it comes across small but in a way seems to make itself bigger. Seems complicated? It's sophisticated! The ornamental painting of the two-story street elevation somehow plays with the discrepancy between the painted heaviness of the Cyclops masonry of the lower level and the repudiation of heaviness at the upper level. As already mentioned: This is witty and not just frolic. If irony has not gone flat in a quarter of a century, we can call it success.

Front facade

Christoph Dietlicher, **Thomas Drack**, and **Andreas Giupponi** founded the **"Neue Werkstatt"** (New Workshop) in 1988. Dietlicher and Giupponi are jewelry designers and silversmiths, Giupponi is also a precision mechanic. Drack is a mechanic (he left the group in 2005). They began with the principle of producing what they designed. They made a positive design statement using a narrow range of tools that enabled them to work with simple machining techniques – cutting sheet metal, bending, folding. Deep drawing, spherical shaping, or pressing did not belong to their repertoire. Curved forms in sheet metal were achieved by bending the sheet along a curved line of holes – the row of holes serving as the geometrical location of the intended shaping. The wall lamp **"Banana"** is an attractive example of a piece employing this technique. ——— In the last several years, the Neue Werkstatt has established itself on the commercial market with a collection of lamps that includes different models for the home and workplace. They are carefully designed and constructed in-house and use the latest fluorescent lighting technology for top energy efficiency. ——— The **Penguin floor lamp (1993)**, selected here as a second example, marks the beginning of this direction in the Neue Werkstatt's work. The lamp is made of an isosceles triangle of sheet metal, bent at a right angle along its central axis. The upper corner is then bent forward, giving the lamp its distinctive form. In the final steps, the mounting link for the lamp socket is tack-welded, the hole for the cable drilled, the sheet metal powder-coated in white, and the electrical fittings mounted. The curvature of the side lines expresses the forces of the machining and is an important part of the formal statement.

FREITAG

The name reveals what this piece of furniture can do: **Shuhkippe** (shoe rack, literally: shoe tilter). It holds ten pairs of shoes in an upright position. It is therefore only 15 centimeters deep, although the average shoe is 30 centimeters long. Designed by **Hans-Peter Weidmann** in **1986** and made of hot dip galvanized sheet steel, it quickly became a classic owing to its character. Frequently copied all over the world, it remains unparalleled in design. Its combination rod / handle opening mechanism with integrated buffer spring distinguishes this design from the rest. ——— Remarkably, a Schuhkippe had already been produced in Switzerland (it was even called the same). Designed by Willhelm Kienzle (see p. 48) in 1951 out of wood, it immediately became a bestseller among living accessories. Kienzle also designed a metal version for the Blattmann metal works (1955). In contrast to Kienzle's design, Weidmann's shoe rack is attached to the wall and does not touch the floor. This is an important break from the fifties, when walls were the untouchable wallpapered realm of the landlord. ——— Weidmann was, at best, unconsciously aware of Kienzle's model. Perhaps Switzerland has a sociocultural constant called "space saving" which acts as a creative catalyst. (Life in the steep mountain valleys, cultivation of the dry stone wall terraces of Tessin, or the thrilling adventure of building row houses are all rather unlike the boundless open prairie.) Japan, the Netherlands, Italy; all are familiar with the problem of limited space, which is an incredibly powerful source of design inspiration. The shoe rack's bold colors are very today: light green, light red, light blue, combined with anthracite, galvanized sheet steel.

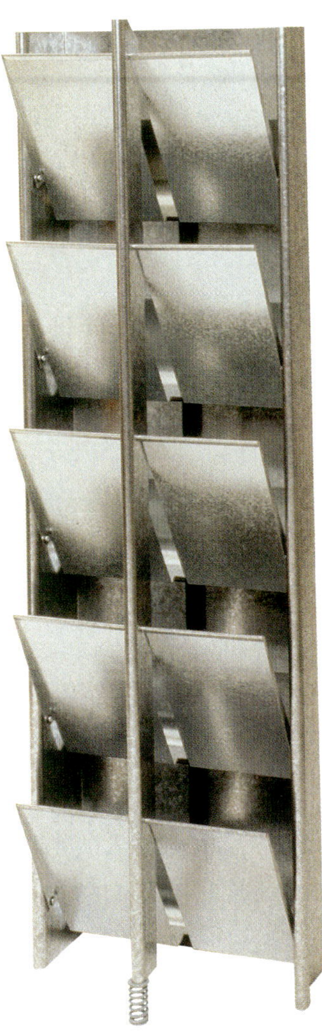

“Typography renders language legible” and “Design is information – it influences a text’s message,” these are two laconic maxims by **Hans-Rudolf Lutz**, whose meaning he developed through his work as a typographer, as an instructor, and as an author. The first sentence, a seemingly innocuous commonplace truism names the primary function of writing in society: It renders that which is spoken as mere ephemeral vacillations in the air. However the entire sentence reads as follows: “Typography renders language legible – to this end there are many possible ways to do so.” This is a momentous revelation that declares the classic conjecture of the division between “content” and “form” as a fallacy. Ultimately, this has been the main focal point of Lutz’s work, herein lies the crux of his oeuvre. His teachings, laid out in his books, remain fresh, meaningful, and relevant even years after his death. These self-published books are incomparably significant compendiums; reading them, it becomes evident that the driving force behind his work was an eminent, inventive, critical passion for visual communication. Books such as “Typo. Basics of Typographic Design” (1987) or “Typoundso” (1996, the first chapter is entitled “The Democratization of Typography”) are highly condensed communication experiences of phenomenal design,

Hans-Rudolf Lutz, poster “Rosa Luxemburg”, 1971

print quality, and materiality. These are books! Lutz significantly contributed to the overall profession of typographical design being understood by the principal exponents of the profession as a co-authorship, rather than solely as a service. This premise is of central importance to Swiss literary culture. Lutz was not the only one who had and conveyed this opinion, yet he did it methodically and uniquely and with great personal charisma. ——— The first example is a performance of an internationally renowned group called **UnknownmiX** featuring the singer Magda Vogel, a long-term project **(1983 to 1992)** aiming to reveal the close relationship between the musical and the visual arts. Lutz considered himself an active participant of this group, involved with the projections of images and texts during the performances, and not simply as a stage designer. The second example is the **Manifesto** "On the Russian Revolution" by **Rosa Luxemburg**. In **1971**, Lutz typeset this book, whereby simply via the differentiation of one font (in this case Antiqua Melior) in "normal" and "semi-bold", the portrait of this German revolutionary becomes visible. Another one of Lutz's common sayings was: "Writing and Image: the transitions are fluid". This is an especially subtle attestation of this.

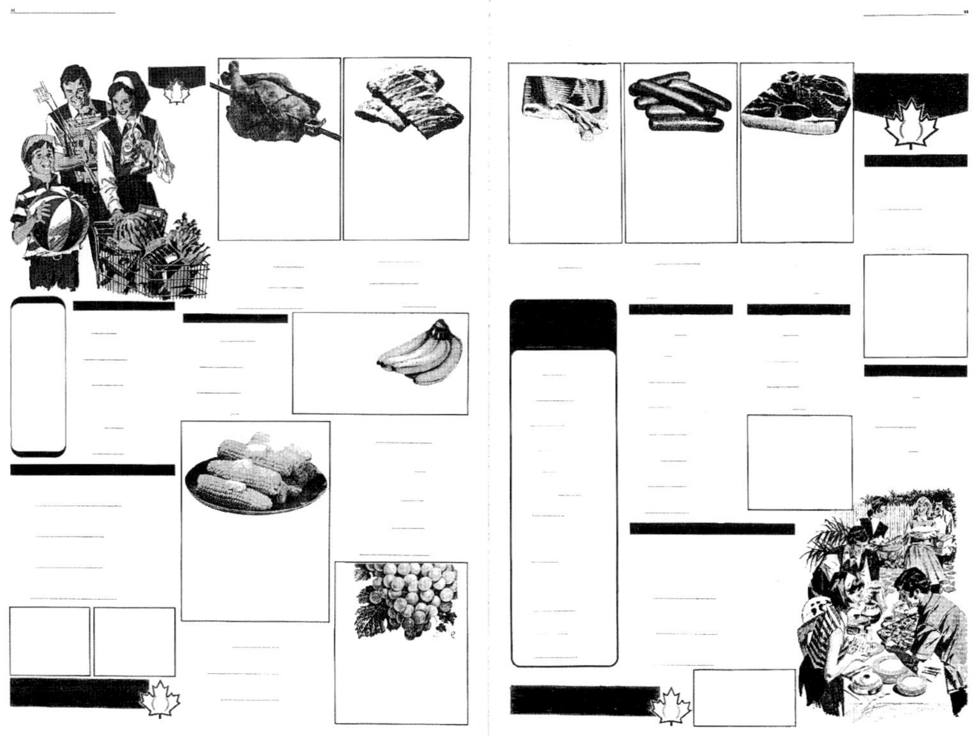

Double page spread from the *Edmonton Journal*, 1978

158/159 Performance by the group UnknownmiX

may spoil your life jungle jive flimmering jungle
whispering jungle jive dangerous thing jungle jive heating
bodies they are slightly moving
slight tiny fresh cold breezy makes them
finding nothing except the own hot wet passionate flesh
mean trap that is a bad joke this dream is
jive flimmering jungle jive whizzering
jungle jive heating sting. Feeling
fresh. Jung
whispering
that is moving too snakes a
jive may spoil your life
back
jungle jive dangerous thing
heating creeping up and
bursting
only that is the
jungle jive
heating sting
air

kakadu

160

Werner Jeker, exhibition poster Man Ray, Lausanne, 1990

Graphic designer **Werner Jeker** has created a life's work of posters as large as it is momentous. He is a founding member of *les Ateliers du Nord* (see page 246) and has been working for many years for cultural institutions such as the Musée des arts décoratifs (today Mu'dac) and the Musée de l'Elysée (both in Lausanne), as well as for theatrical institutions and private industry. One strength of his many works is the impact they make in public spaces. Similar to Armin Hofmann and other designers, he includes as one of his design criteria that his posters stand out from the probably clamorous patterns of the surrounding ones. That is why he often consciously chooses only a limited spectrum of colors for his works. An earmark of his posters is the use of details and a clear division of area thanks to the use of clear contrasts in color and brightness. ——— In this **1981** poster for **"Wohnshop"**, a furniture dealer in Lausanne, Jeker uses the same distinctive angle he had previously introduced as an emblem for the Theater Lausanne-Vidy, building considerable tension between the very pliable form of the larger-than-life chair and this rigid element in the foreground of the image. The poster develops its full three dimensional character precisely because of this attempt to suppress it. In other cases, Jeker reacts flexibly to the inferior quality of the prints he works with, allowing a grid to become visible. In the **Man Ray poster** for the **1990** exhibition at the **Musée de l'Elysée**, he gives voice to this art photographer, who attempted to determine the best detail of a photo by folding his prints. Jeker makes it immediately clear that creating art is a self-critical process long before the final work is finished.

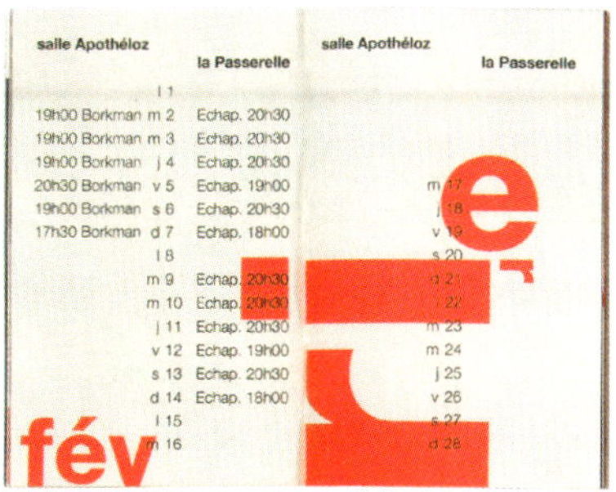

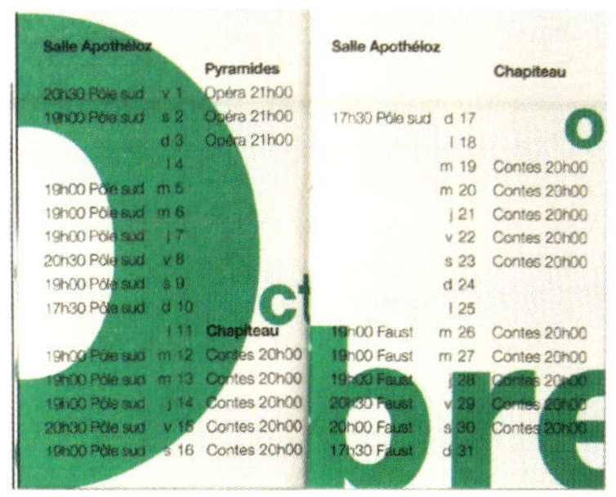

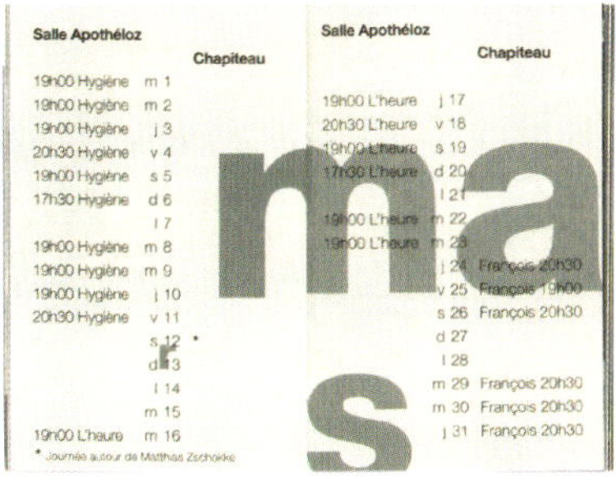

Program schedule for the Théâtre du Vidy, Lausanne, 1991

162

Werner Jeker, poster for "Wohnshop", Lausanne, 1991

Beat Frank, originally a visual designer, moved into three-dimensional design only in the 1980s. In 1984, he founded the **Atelier Vorsprung** (head start in English), which he has run on his own with the exception of the period 1986 to 1990, when he collaborated with Andreas Lehmann. When he talks about his mission, he speaks of "simplicity which has gone through complicatedness." The form is to be created less by human will than through the immanent logic of its own laws. Accordingly, Frank is not concerned with creating superficial elegance, rather with a deeper reason for a form. If the form – in his words – "makes itself," it will not only be perceived passively, but will actually be created anew by being perceived and in the act of perception. ——— Believing in "animated matter", Beat Frank is interested in joined objects, in other words, connections in a manner logical to the form. He hardly ever uses screws or the like. The elements are just fitted together, and the logic of their assembly (literally: what holds them together) is their central characteristic. ——— His **library shelf** is a hollow walk-in unit. It consists of vertical planes of pickled sheet metal 6 mm thick and horizontal shelves made of birch plywood (or optionally MDF), for a total of 40 running meters of shelves. The sheet metal elements are varnished in matt gray. They are laser-cut, combining archaic joining techniques with highly developed technology. The image shows two such units in the studio of a well-known writer near Zurich.

The folding grille is an ingenious object. Although we can see how it works, it remains somehow mysterious. Its form is flexible although its individual parts are not. Pushed aside in an entranceway, for example, we overlook it; when it is closed, we can still see the way in, but now there's also an imperious barrier.

——— We remember a trivet protecting the table from the hot casserole dish at family dinners. In **1990**, **Kurt Thut** applied the folding principle to the **Scherenbett** (literally: scissors bed): simply supported wooden slats which unfold to bring forth an area of variable size, supporting either a single mattress when pushed together or a double mattress when pulled apart. An

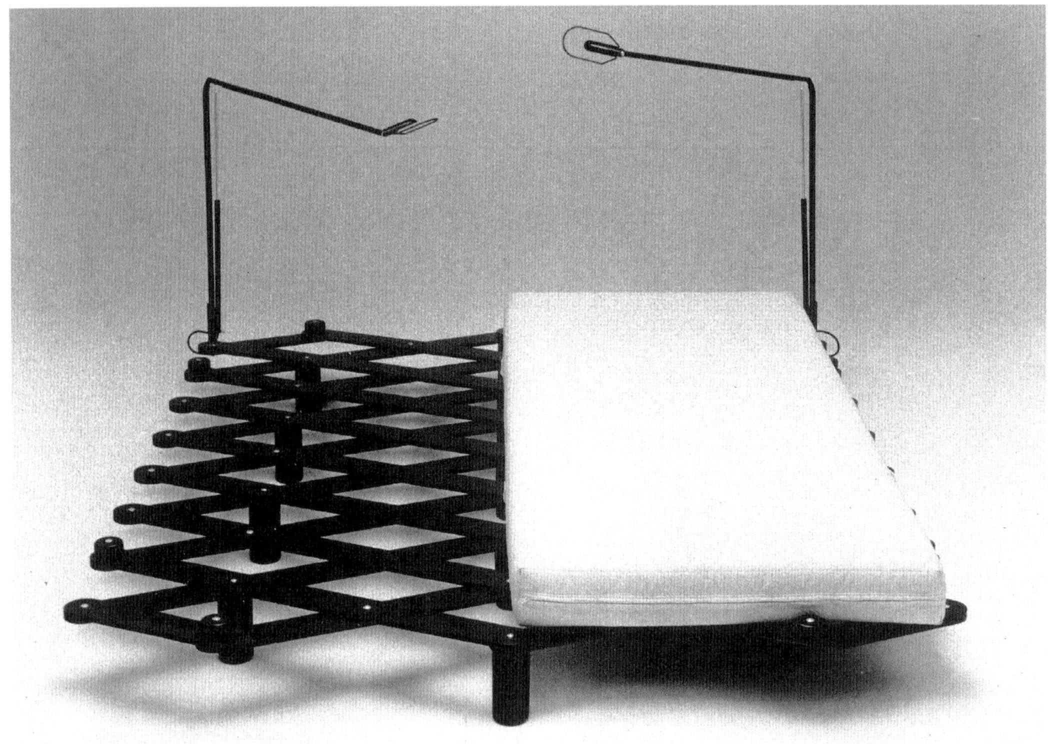

idea so convincing that one wonders why it took so long for it to be put to this purpose. Theoretically, we could unfold the grille even further (larger than its planned maximum size), creating a structure approximately three meters long and half a meter wide. We don't have to go this far, but we could. Wooden bumpers at the joints prevent the mattress from sliding. Benjamin Thut's **Lifto-Leuchte** lamps can also be attached to them. Kurt Thut won "die gute form" ("the good design") award in 1955 for a low, elegant sideboard with metal sliding doors; a piece which, reduced to the max, could have been designed today–designer Kurt Thut, a time-traveler.

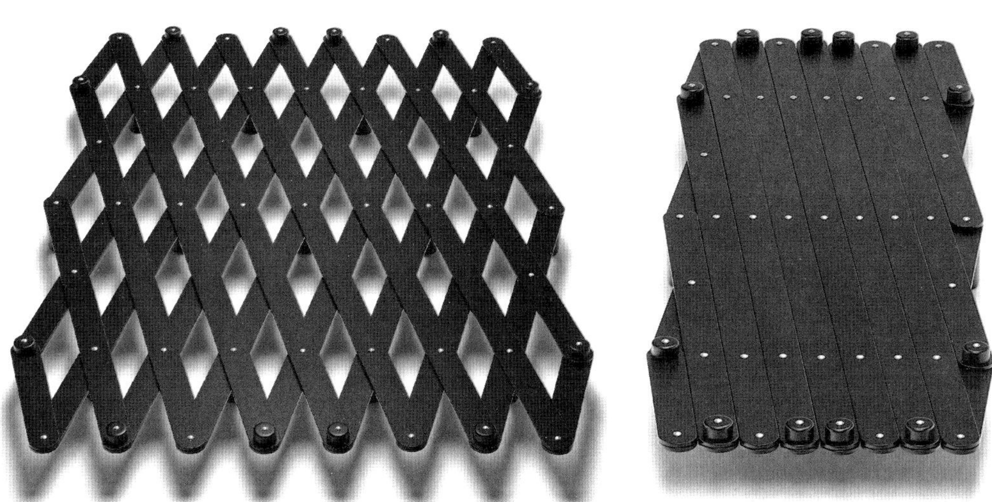

When you hear of a table five meters long, you tend to think of perhaps three tables pushed together to seat a large gathering for a special occasion. Maybe the tablecloths will have to conceal the differences between the individual tables? Nothing against that. However, the **table** in question here is five meters long in and of itself, and its legs are placed at the outermost corners. If not for this length, one might speak of the archetypal table. But: this length! How can it work? Why doesn't the table sag? A good craftsperson would have made a table five meters long with cantilevered ends, for example, so that their weight would take pressure off the middle section; in this case the span between the legs would be only, say, three meters. But here, the legs are fully five meters apart. Are designer **Silvio Schmed** and manufacturer **Ph. Oswald**, who developed the model in **2003**, inferior artisans, or are they magicians?——— Because in this case, it's about something other than sheer pragmatism, namely finesse. The tabletop has a light honeycomb core and is slightly cambered. The vertex (in the middle) is approximately one centimeter higher than the ends, that is, 0.2 % of the table's 500 cm length. These 0.2 % are sufficient for a thin layer of synthetic resin, bonded to the entire underside, to provide the necessary tensile strength to prevent the tabletop from sagging. As with bridges – and the table is a bridge – prestressing helps to work against gravity. However, the great, internal, and invisible feat is how the legs are fixed to the honeycomb core. At this point, it is no longer just fun to look at; this is taming of matter.

Poster designer **Ruedi Wyss**' works have dealt with contemporary music for more than 20 years. His **posters** interact with their subjects on several levels in terms of their design and layout. They are forceful in their graphic idiom and always promise eventful music. Viewed in the order of their creation, they also document a conscious "interplay" between the technical processes available and the status of the client at a particular time: from photocopying to offset printing, then on to silk-screen printing; from Repromaster and Letraset-screen / grid gels to the computer screen; from newspaper cutouts of agency photos to the reuse of images or quotations from famous artists or photographers (Lissitzky, Schwitters, Man Ray etc.). In his early works for **concerts** in the **Schweizerbund Restaurant**, Wyss occasionally utilized children's rubber stamps depicting letters of the alphabet, or steel needles which he spread across the surface of the photocopying machine. If the machine did not work according to his wishes and produced strange copies (even if for other reasons than the needles), Wyss reacted spontaneously as a jazz musician would, and continued his piece with its transformed basis. Like the music for which they were designed, the posters do not aim to appear calculated down to the last detail, but preserve in their appearance the air of their creation. Wyss used block letters of the Eurostyle typeface well-suited to be layered, disarranged, cut, tilted, or turned upside down from 1985 for the B4-format posters advertising the annual **"Taktlos" Festival** (which he personally co-founded). The second **festival series**, "**Tonart**", has a different musical emphasis, which is expressed accordingly in the idiom of the posters.

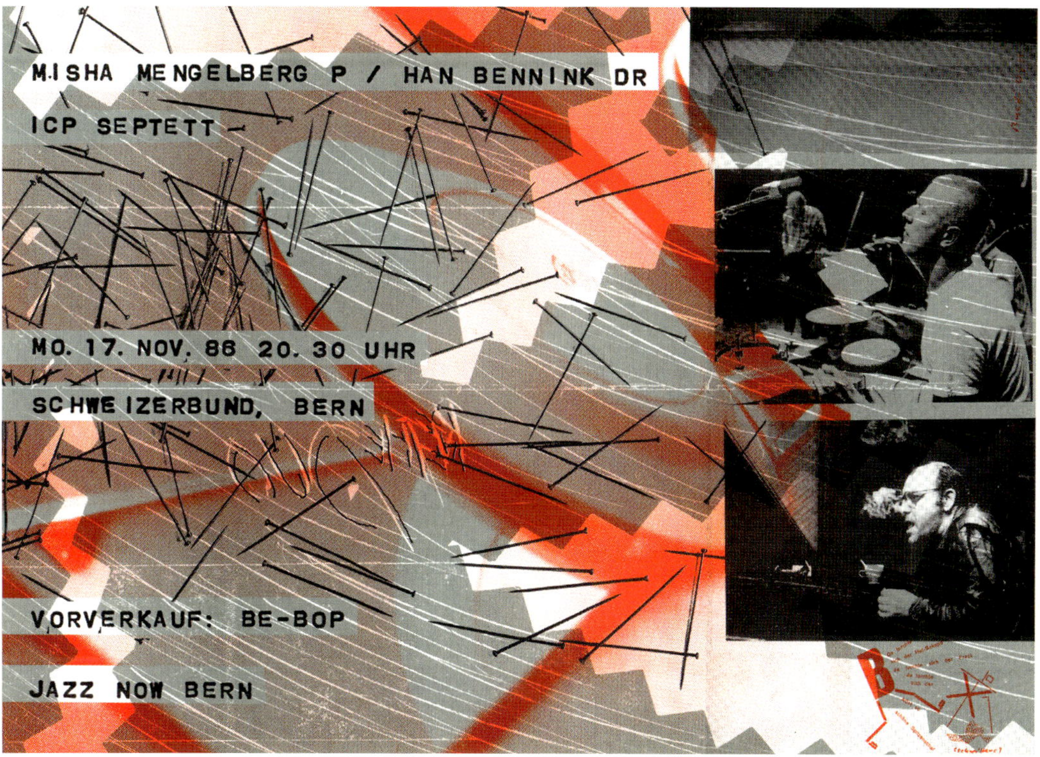

Ruedi Wyss, concert poster "Han Bennink/Misha Mengelberg", Bern, 1986

Ruedi Wyss, festival poster “Taktlos”, (Bern, Basel, Zurich), 1996

Ruedi Wyss, festival poster “Tonart”, Bern, 1989

Georg Staehelin is an eminent conceptualist. He pursues the form to fit the content, that is to say: the form of the content. What the content is becomes completely clear only after its form has been found. The poster for the **1996** exhibit **"Netto. Nichts als Inhalt"** **(Museum für Gestaltung, Basel)** is a literally striking example of this maxim. Staehelin again and again creates both a riddle and its answer. His works are an expression of essentialization, manifestations of the search for personal realization – the process of realization itself and an overture of communication to the public. The impatient sometimes say one must be able to grasp a poster in the space of two seconds. Dear people, Staehelin requires more of your time, but it is overtime that is well invested (to stay with the terminology); Staehelin draws the observer into the image creation process, where one forgets to count the seconds. As a designer, he searches for the key to unlock the contents of a topic and leaves it within the observer's reach. It is the observer's job to take this key, put it in the lock and turn it. ——— Staehelin is a student of Armin Hofmann – one of those students who does not formally or stylistically continue Hofmann's approach, but rather carries on his manner of thinking. Whether one comes across Staehelin's work for the furniture manufacturer Lehni, for Wohnbedarf Zürich, or for the St. Gallen fashion label Akris – his prints always encompass the utmost of exactitude. They have enormous powers of persuasion, are printed on heavy quality paper, are rhythmically structured, accentuated by a clear arrangement of form, color, and space, boasting white whites, red reds, and black blacks.

Georg Staehelin, poster triptych for a double exhibition, Basel/Weil am Rhein, 1997

171

Georg Staehelin, exhibition poster "z. B. Schuhe", Zurich, 1988

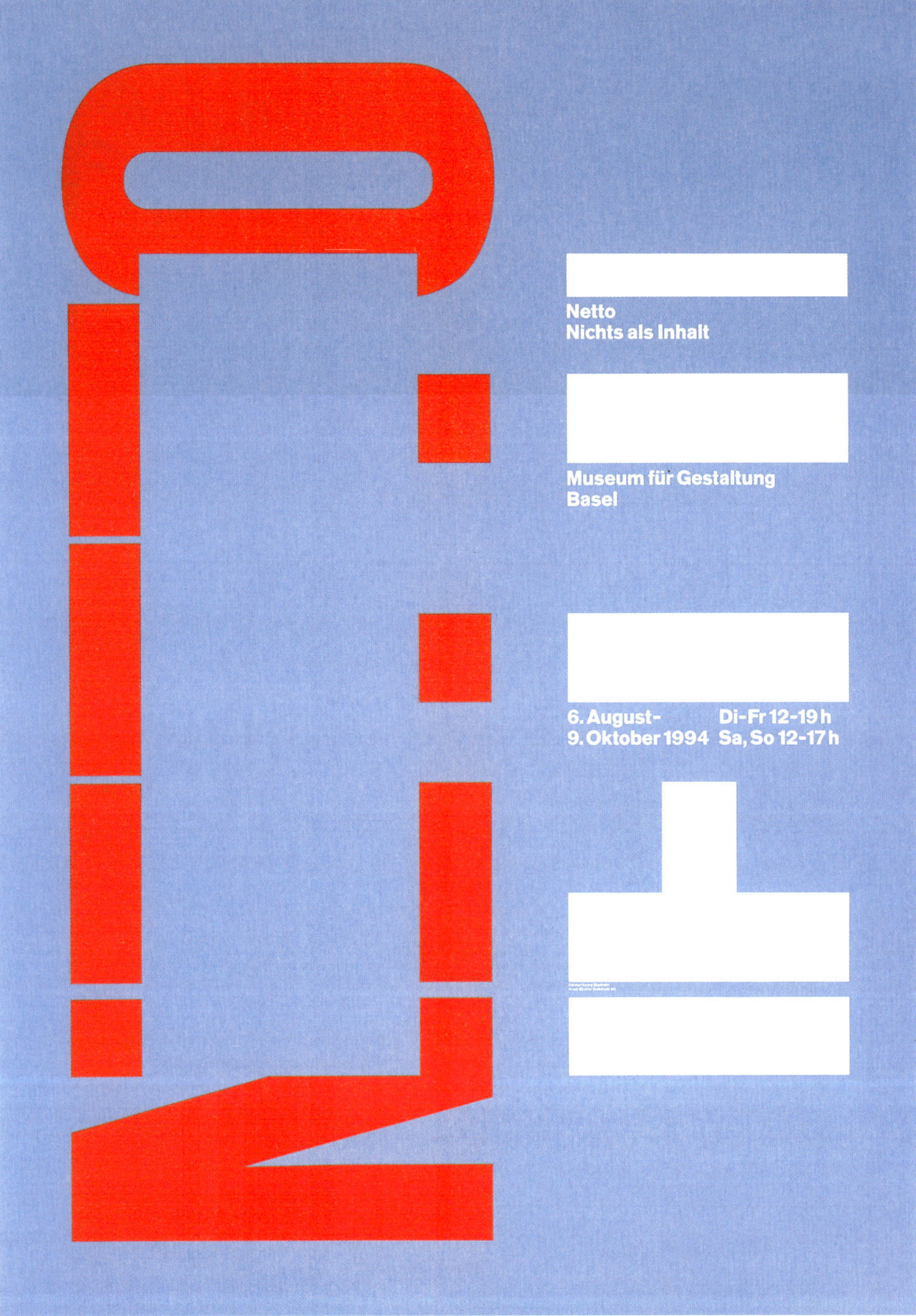

Georg Staehelin, exhibition poster “Netto. Nichts als Inhalt”, Basel, 1996

1990–2006
POSITIONS / FIN DE SÉRIE

The current era has not yet been named, and that is a good thing. We make do with terms like “the age of globalization” or “information society”. In a chronicle of events, the beginning of this period coincides with the dissolution of the Warsaw Pact and with the revolutionizing of how we work as a result of the personal computer and the internet, which developed as a consequence.

The most striking difference in comparison with the phase after 1950 is that the idea of linear progress is losing credibility. Each individual life is a project with an unknown end result – that is nothing new – but today, we feel that’s okay. Existential insecurity is no longer a vexation, but rather a motive to define oneself as a subject. That is why developing a position is an important goal in what we call creative work (The inflationary use of the adjective “creative” proves the statement made above: the staff member who sat at an oak desk wearing an apron fifty years ago as a clerk would consider himself a member of a “creative team” today).

What distinguishes the current times can, as always, be viewed from three angles: critically, affirmatively, or declaratively.

Speaking *critically,* our age has become consumerist to an unprecedented degree. This shows our uninhibited egoism, among other things. We only live once, and only by our own efforts. One is not merely subject to the play of forces, one is a part of it. What, 150 years ago, was considered a (happy) destiny which people were to forge themselves is today one’s “market position”.

Declaratively, we observe how new urban customs have developed and spread deep into rural areas. Nowadays, young people move between cities the way a generation earlier, they would have moved between neighborhoods. The club scene, cell phones, GPS, and a discriminating culture of entertainment lead them to mediate the appearance of a real lifestyle akin to what their grandparents imagined a well-paid life of leisure in America to be like fifty years ago.

The principle of variety guides the regime with a certain charm, but at the same time with iron relentlessness. One’s own individual view of the world becomes a topic and puts the image thus sketched of the surrounding world (the environment) up for grabs. What must be considered egoism when viewed critically could also be called individualism. Is that sophistic hairsplitting? No, for it may be that the dividing line between the two terms is simply the difference between passive consumption and active creation.

Spoken *affirmatively* and with a view to the theme of this book, personal authorship is a signature of this epoch. It is no longer possible to achieve a profile as a designer merely by understanding the service to be provided. Especially when it comes to visual design, personal authorship has been declared an outgrowth of the youth movement. Considerable evidence supports this thesis, but this explanation is also linked to the desire of an economic system in which basic needs have been satiated for an attractive appearance of its own. Achieving a strong identity presupposes differentiation vis-à-vis one's competition, and vice versa–why: Identity is differentiation. Graphic designers, architects, and industrial designers are the most important partners in this quest.

In **graphic design**, it is new fonts, new papers, new folds, perforations, and ideas of all kinds that unmask the objectivity of "black on white" as a crude simplification. It is at least as much about the form of conveying the content, about a "climate" and an associative dimension as it is about the content itself.

Today's designers' ability to define a particular concept and to defend it self-confidently is remarkable. This must also have to do with the limitless possibilities provided by the tools available. When the computer has hundreds of typefaces and ten thousands of colors at the ready, only proactive selection can help, and the foundation for making such choices is the capability to reflect one's own actions critically.

A related insight is that the mere existence of an object (a building, an installation, a design object, a visual work) is not sufficient to attract the amount of attention intended in the overcrowding of simultaneous events. Publicity and media attention are necessary, the critics must take notice, in short: It takes a discourse.

It appears evident that "ingenuity" plays a major role in this process of cultural communication. It includes the public and turns viewers into fellow players and fellow explorers. The difference between now and the years before 1990, when the understanding of form was still focused much more strongly on the tangibly material, is meaningful. Today, form is rather something that comes into being through the act of perception. Perception is an integral part of the form, the public has grown out of its role as applauding backdrop and has itself become an agent (licensed to act). The public's attention to this position is remarkable, as the Swiss National Exhibition Expo.02, for instance, showed.

In **architecture,** at first glance, the focus is directly on the object, in contrast to the preceding era, the question of context appears to be subordinated again. But this appearance is deceiving. In contrast to the 1980s, the context has simply been redefined: it is no longer merely the matter surrounding the building, but also the "semiosphere" and the imagined user's dynamic mind map. To give an example: The terminus of a regional railroad line near Bern is not just located in the built fabric of a village. Rather, whoever sets foot in the station is already to some extent mentally in Bern. This cerebral dimension is thematized within the building's design. Architecture no longer needs to justify itself as a medium; the question is at which level one can play this role.

Fragmentation, the kaleidoscopic, variety, and the attention-getter are means to prevent losing one's orientation amidst confusion. Or are they even a means necessary to afford lack of clarity? A hundred and fifty years of the railroad, a hundred years of film, fifty years of flying and television, as well as ten years of the internet have changed the apparatus of perception profoundly. When everything is in motion, everything happens in motion. Then clarity is part of the equation, because it adds justification and stability to the volatility of the images.

The bicycle parts company Shimano commissioned **Hannes Wettstein** in 1996 to develop new concepts for the bicycle. This catalyzed him to think about a fundamentally new bicycle configuration. He wanted to find an alternative to the legendary diamond frame which had given the bicycle its form for a century, since around 1895. A major disadvantage of this frame is the need for complicated guide pulleys for brake and derailleur cables and the ensuing obstacles to their reliable operation. Insufficient torsion resistance in the women's bicycle is a further weakness. ——— Based on this analysis, Wettstein redefined the geometry of the bicycle frame in **1998**. While the primary triangular element of the diamond frame, to which the seat stay and rear fork end are attached, consists of the top tube, the down tube, and the seat tube, the **EST-Rad** (EST bike) features a different configuration. The top tube leads diagonally to the rear hub, forming a trapezoid together with the seat tube, down tube, and head tube. This makes it possible to conduct the cables in a direct line. Since the head tube is longer than in the diamond frame, greater height adjustment of the handlebars is possible, and thus a larger range of individually optimal seating positions. Wettstein patented this compact frame in 1998. Since then his company **zed** has been expanding its collection. The EST line offers bicycles from 20 to 26 inches in different styles. What they have in common is their unpretentious look; they do without the usual decorative elements so as not to diminish the beauty of the new compact frame's economy.

The mountain village of **Vrin**, at the end of the Lumnezia valley (Grisons canton), is the home of Romansh native **Gion A. Caminada**, whose work tells an unusual (and touching) story. His main frame of reference remains Vrin, even when he builds elsewhere. A trained carpenter, Caminada's first work was for the people of his village: a barn here, a house there, then vice versa, then a house once again. Here you can see **two barns**, a **drying room** for meat, the churchyard **mortuary** (stiva de morts), and the **community hall** – all built between **1995** and **2002**. The visible joining of robust, planar walls and roof to create their compact volumes gives these buildings a direct and striking character. In the *Strickbau* tradition, on which Caminada bases his architecture as a matter of course, the milled timbers that make up the walls are dovetailed and fitted at the corners, their protruding ends giving clear exterior expression to the interior walls. The "interwoven" wall members give the timber frame

Strickbau dwellings the look of almost archetypal houses in which humans or animals can find refuge from the wind and cold. In these houses, there is a legible congruence between inside and outside. But new *Strickbau* houses can often be found – what makes Caminada's buildings special? ——— It is his engagement with the tradition of construction in the mountains of Grisons, with building in wood and stone. Caminada consciously refers to this tradition anew, albeit with a surprising interpretation. "Tradition and innovation", an attractive idea – what a shame that it so often remains an empty slogan. For Caminada, it is much more than a slogan. He challenges tradition, puts it through its paces, rejuvenates it. He makes tradition conscious of its own worth. This is his great achievement. One might think that traditional Strickbau construction leaves no room for modern themes: the flowing spaces, the separation of supporting structure from the division of space, the walls that

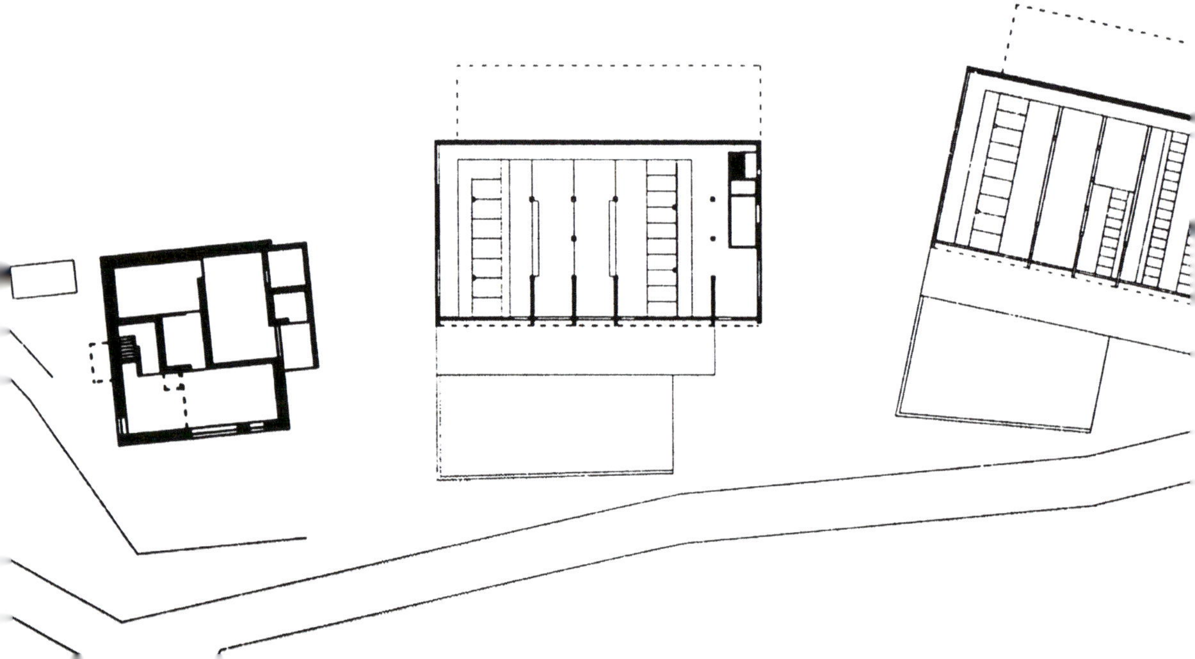

dissolve into glass. And certainly it leaves no room for a flat roof, a building element which is not native to Grisons anyway. What Caminada does, however, shows us how superficially we tend to understand such attributes. ——
In his work, he is able to reflect these themes, not just rhetorically, but as materials we can experience with all our senses. A window is more than a hole in the wall, it is often its own small space, suspended in the weave between inside and outside. A zone is more than a functional plane, it has its very own place in the division of space. The ceiling is more than just a fifth wall: It is an authority, an organizing factor. These are old themes in Grisons' buildings, themes which generally receive cursory attention at best in the design of new buildings. Usually, they are completely overlooked amid the impudent rhetoric about rough-stuccoed walls and hand-hewn illusion. Caminada seeks out these old themes, apprehends them anew, and exposes them to view.

182/183 Part of the village of Vrin from the south, on the right the group of buildings with the drying room for meat and the stables

Interior of the multi-purpose hall (with Jürg Conzett)

Church with the Stiva de morts/mortuary

Floor plans of the drying room for meat and the stables

South elevation of the drying room for meat

Children like to use car tires as inner-tubes when swimming (this is common from the era when car tires were still tubeless). Blow-up furniture has been an item ever since Buckminster Fuller in 1930 and had a come-back later in the sixties in Italy and England: Jonathan de Pas and Archigram, and later Tom Dixon. **Sabine Leuthold** and **Jacqueline Lalive d'Epinay** (Ho-La was the name of their Atelier) were certainly not the first to conceive of this idea in **2000**, yet they took it more seriously than just a "sudden impulse", they developed a product based on this idea and called it **"Amöbe"**. ——— Particularly the decision to tie the tires together with a belt turned their idea into a piece: furniture, a chaise longe, a "love seat". The belt material was specifically chosen so that the buckle would be well supported, and tubes were sought that neither leave marks on clothing, nor smell like pitchy rubber. They found these accessories and materials and they gently rolled the Amöbe out of the workshop of their imagination and on to the floor for product development. They even developed a line utilizing colorful rubber. Some cloth, a blanket, or a piece of fur turns the "Amöbe" into a cultivated objet d'art for design lovers and those with an eye for details.

It would be interesting to write a story of success reached by failure, of unexpected accomplishment by error. Those who think one-dimensionally will think only “failure” and avoid it the next time. Those who allow themselves to think out of the box might pay attention. This is true of **Ralph Schraivogel**. He refers to the **poster** for the **Cinemafrica 1991** film festival (Film Podium Zurich) as a decisive turning point. It begins with a trial sample gone wrong when he tries out his new Repromaster. For some reason, Newton’s rings show up on the sample, making it unusable. Schraivogel saw that he had come across something new. For the 1989 Cinemafrica poster he had drawn a ze bra pattern by hand. Now he wanted to know what was what and intentionally produced Newton’s rings, which he then manipulated using tiny rub-on letter transfers of the word “cinemafrica”. Schraivogel then enlarged this small original (only a couple of centimeters in length) to standard draft size. He transformed the left side to counteract the right side (black-white inversion), thus condensing the design even further. ______

Schraivogel’s posters created by means of an analogous technology (up until 2000) repeatedly, almost mysteriously, express this gestus of layering; going from layer to a space of image and meaning. This applies to his poster on **Henry van de Velde (1993)** as well. After 2001, Schraivogel works on the computer. He resists the seduction of this tool (which could provide him with effects of the kind he previously was only able to achieve though arduous searching) with deceptive ease; from this point on he works only with surface planes. The posters for the **Paul Newman** retrospective **(2001)** and for the exhibit on **British Graphic Design (2006)** are very different, nevertheless they are also very powerful works with their intelligent choice of methods, their incisiveness, and their sure use of planes, colors and proportions. To me, they always also seem like a declaration of love for the standard paper size B4.

188

Ralf Schraivogel, film poster "Cinemafrica", Zurich, 1991

Ralf Schraivogel, exhibition poster “Henry van de Velde”, Zurich, 1993

JULI/AUGUST 01

PAUL

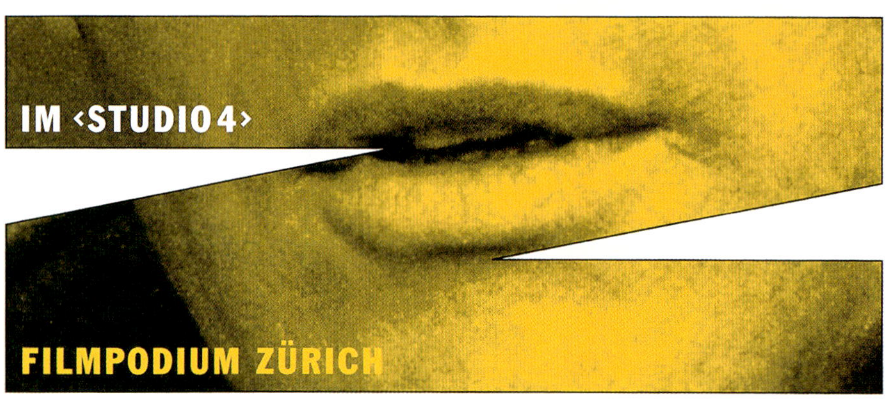

Ralf Schraivogel, film poster “Paul Newman”, Zurich, 2001

Ralf Schraivogel, exhibition poster “Communicate”, Zurich, 2006

This small theater in a courtyard in the **St. Alban-Vorstadt quarte**r of **Basel** was established in **1973** by Ruth Oswalt and Gerd Imbsweiler in a former workshop, and has been run by them ever since. They have been honored with the highest award in Swiss theater – the Hans-Reinhart-Ring. Their repertoire consists of performances for children and youth as well as for adults. The original name "Spilkischte" (toy box, literally: play box) was rather one-sided and was changed to "Vorstadttheater" after renovation. The remodeling of the theater on the occasion of its 25th anniversary in 1999, carried out by architect **Thomas Schregenberger** (with Lorenz Peter), had to be "staged" with a minimal budget of 110,000 Swiss Francs.

——— A passageway leads from the street to a trapezoidal courtyard that borders the small theater at the back; large letters spelling "Theater"

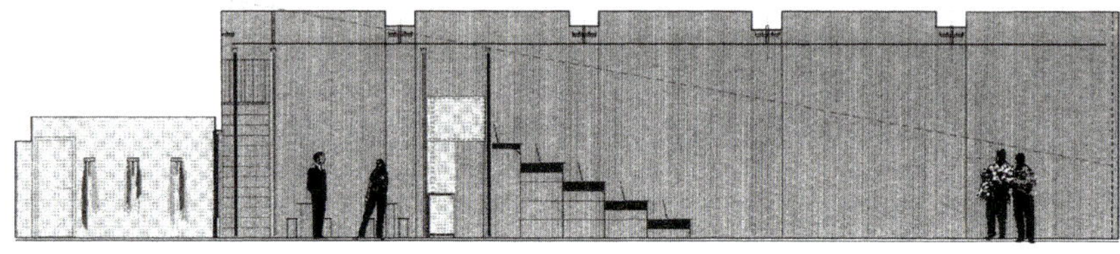

fill out the part of the facade visible from the axis of approach. These slender initials resemble a geometric ornament whose graphic black and white cover the entrance to the theater and more. The open entrance bids the public enter with an almost ostentatious emphasis. Opening outwards and painted red with glass insets, the wings of the double door are like a footman calling “Step right up, ladies and gentlemen!” The lobby, painted the same red color and boasting cubic ceiling lamps, glows from afar, leading playgoers from the quotidian world to the world of the theater. Movable platforms and a curtain rigging system in the auditorium allow for a variety of uses. Peter Brooks’s concepts of “empty space” and of theater at eye level with the audience greatly helped the theater managers remain certain that great theater is possible with little material expenditure.

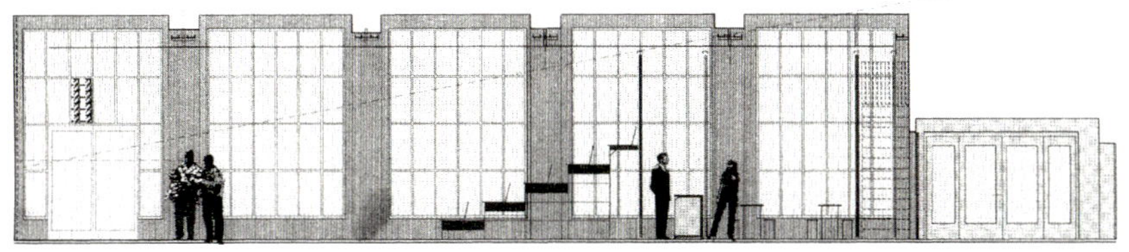

Mount **Niesen** (2,336 meters above sea-level), located above Spiez in the Bernese uplands became known as a monument due to Paul Klee's early watercolor (1915) portraying it as a blue pyramid. The old **alpine chalet** built in 1856 near its peak, received a generous **extension** in **2002** from the Bernese architects **Aebi & Vincent**. The annex is conceived of as an overhanging horizontal platform, where the path from the lower set alpine station of the Niesenbahn funicular ends. It forms a horizontal cut into the pyramid. Unfortunately this pathway is (still?) not yet designed as it was originally conceived and proposed in the competition design: as a zigzagged course, where each turning point was also to be built as a small wooden platform. At night, luminaries set on the path were to provide it visibility, lighting it up like a "constellation" – this draws to mind Bruno Taut's "Alpine Architecture".

The annex is an expansion of and is subordinate to the existing alpine chalet.

It extends to the side and out to the front towards Lake Thun. The steel skeleton construction rafts were transported to the site via helicopter and were installed in little more than a day. As a supporting structure, the annex is quite a feat of construction – able to withstand wind velocities up to 240 km per hour and up to 3 meter-thick layers of snow. The reinforcement against occasionally enormous horizontal forces was astonishingly nevertheless achieved without the use of diagonal lattices. The profile of the roof rim is designed so that a storm can not lift and dismantle the roof, rather it even has the opposite effect of a downward force. The painted blue ceiling of the main room is not an allusion to the cloudless sky, rather the architects' purpose was to create an optical connection to Mount Niederhorn on the other side. Intentionally or not, through this reference they enter into a conversation with Paul Klee about "the mountain".

The **compact disc** had a tough time winning over all those who had come to love the generous format of the long-playing record. An LP cover was often its own little world in the space of one square foot, catering directly to the taste of its target audience. In contrast, for years the CD was negligible, at least in terms of its visual and tactile design. Static-free bytes were enshrined in an apparently sterile acrylic jewel case less than a quarter of the size of a LP cover. The loss of sensual and artistic appeal was painful. **Daniel Volkart** is one of the pioneers who found a compelling way to handle the new CD format early on. His design for the **CD cover** of Mani Neumeier and Luigi Archetti's **1993 "Tiere der Nacht"** already took into account all the elements which

other designers at the time tended to carelessly throw together: the label (RecRec), the bar code, all typographical elements, the jewel case tray which holds the silver disc, and the CD itself. The thin cardboard booklet and the CD form an unusually homogeneous whole. This type of design, characterized by differentiated attention to a specific subject, was a convincing gesture of opposition to the aggressively indifferent standardization of the CD format. Music labels have learned quite a bit over the years, but much of CD design still seems to be accidental; often the design seems to have been planned at a different scale from the finished product. Encountering harmony such as we find here is still a stroke of luck.

The early-20th-century building in which the intervention described here took place in **1998** has a lively cinematic past. This is the location of the "Union" movie theater, later renamed "Modern". In the years since the 1960's when many cinemas closed their doors forever, it housed a billiards saloon. A group of film buffs found a new venue for their film programming passion here on Neugasse in **Zurich** after a long period of making do with provisional locations. For economic reasons, they needed two small screening rooms in place of the one spacious original one that seated three hundred. The architecture firms **Meili & Peter** (Marcel Meili, Markus Peter) and **Staufer & Hasler** (Astrid Staufer, Thomas Hasler) collaborated on developing and realizing the concept which enables using the original projectionist's booth of the "Union" movie theater to show films in both screening rooms at the same time. The screening rooms are different sizes and are laid out at a right angle to one another; the lobby with the ticket booth and the bar is in between. This arrangement required that projecting films in the larger

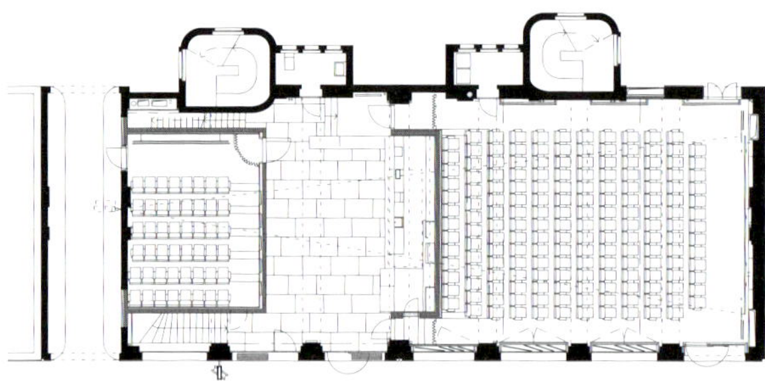

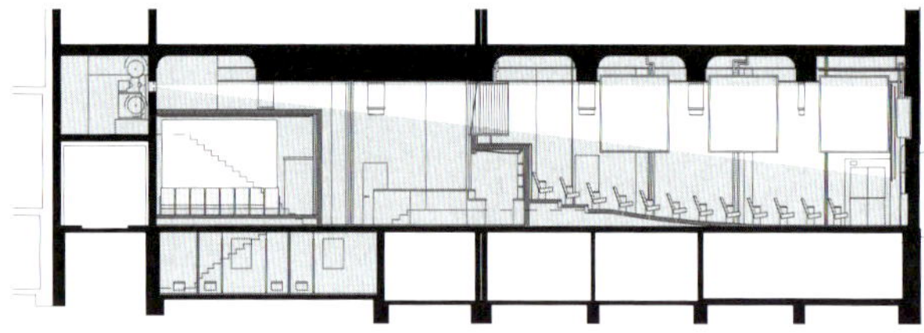

Floor plan and section

Test screening

screening room through the lit public space of the lobby proved practicable. However, this idea contradicts the usual expectation that film projection is possible only in the dark, and that the projection beam would be compromised by the light in the lobby. But according to the laws of physics, this is not the case. The lobby with the bar, where moviegoers get together for a cup of coffee or a drink before or after viewing a film, is thus also the meeting point of the objective laws of physics, people's empirical expectations, and the correction of those expectations in practice – in short, the location of an urban awareness of life. Everyone there takes note at least peripherally of the flickering beam of light, its reflection on the dark pane of glass, and the diffusely reflected goings-on in the film on the opposite wall. The name **"Riff-Raff"** was selected for the movie theater and references Ken Loach's film of the same title, in order to express double programming. The building is exceptional in the choice of materials and its color concept, too. This is a place where film is made visible as an apparatus.

The movie theater in operation

FRED FRITH
STEP ACROSS
THE BORDER

MUSIC FOR THE FILM BY
NICOLAS HUMBERT AND
WERNER PENZEL

Peter Bäder usually works as an illustrator – for the comic magazine Strapazin, for example. He also designed several album covers for Swiss (RecRec) and Canadian record labels. In this case, we're looking at the album cover of the soundtrack from the wonderful **film *Step Across the Border*** (with Fred Frith and Iva Bittová, among others) by **Nicolas Humbert** and **Werner Penzel**. The **1989** black-and-white film, with its intense images and humorous, touching music – the music doesn't accompany the film, rather the film is a visualization of unmistakable musical creativity – is signified on the album cover by 16 fields (8 each on the front and back). Blank white spaces form fragments of letters, some letters are black, all stand upright – run – dance, in different sizes, punch holes in the photographs, run through the picture, whereby at the boundaries of the image fields there are no clear encroachments by the letters, only implicit ones. The cut-out areas occasionally almost form positive spaces. The "step across the border" which the film title calls for (or holds the promise of) is almost but not quite taken visually by the letters and the image fields – it exists as a possibility. On the double sided informative sleeve insert, Bäder changed the concept slightly, as he also did when he later transferred it to CD format. The latter is not simply a smaller-scale version of the long-playing album format, but a further modification of the concept.

The heavily trafficked **Mont-Blanc Bridge** in **Geneva**, which traverses the Rhone just as it leaves Lake Geneva, can, since **2001**, be crossed under by use of a **pedestrian bridge.** The architecture firm **BMV** (Hani Buri, Olivier Morand, Nicolas Vaucher) designed this floating steel structure, partially filled with poured concrete. This "passage flottant" is located under the midpoint of the first span and is connected to the right bank by access gangways. As in other cities built where lake and river meet (Lucerne, Zurich), the Mont-Blanc Bridge is also quite low. Due to seasonal fluctuations of a full meter in the lake's level, a floating construction was the natural response to the problem. In order to ensure enough headroom under the Mont-Blanc Bridge, extra weight had to be added to give the structure a draft of around 40 cm. For this reason, the steel construction – 60 meters long and 6 meters

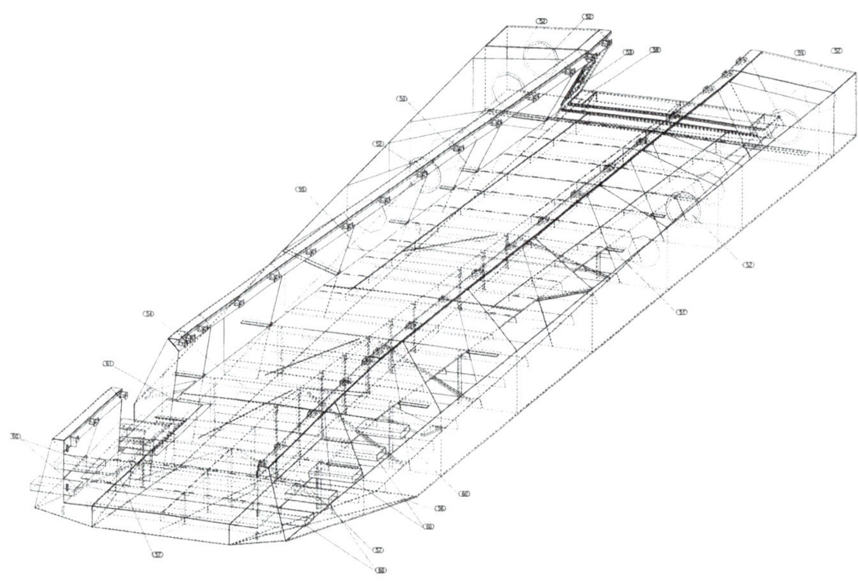

Photographs during construction

Design drawing

The elevated walkway in place

across at its widest point – was partially filled with poured concrete.

BMV gave the footbridge a form similar to a broad laden barge. Their structure, however, tapers at both ends, where it also rises out of the water. Light, flexible, simply supported access bridges connect the footbridge to the quay walls. The deck slopes downwards from both ends, making the submersion in water a sensory experience. At the point where the floating body is widest (and therefore seems 'heaviest'), the path itself is narrowest, while at the tapered ends the deck is widest. The gradual fall and rise of the path is thus accompanied by a continual change in the bridge's cross-section, which is expressed through the successive "twisting" of the flanks. The lighting is integrated into this movement, immersing the path in a reddish light at night and modeling the bridge's forms.

Children and their parents welcomed the new **toddlers' pool** of **Zurich's Mythenquai swimming pool** enthusiastically in **2004**. Replacing the previous, simple flat pool with a gently sloping floor, architects **Haerle/Hubacher** (Christoph Haerle, Sabina Hubacher) designed a multi-faceted water landscape. Its perimeter is a concrete rectangle laid flat in the lawn with diverse articulations on the inside. A waist-high green wall made of finely pored concrete connects it to the small buildings that close the grounds of the swimming pool off from the street. ——— The design's special achievement lies in how it connects subtle geometry with direct experience of the elements. The water flows and moves, and at intervals it streams like

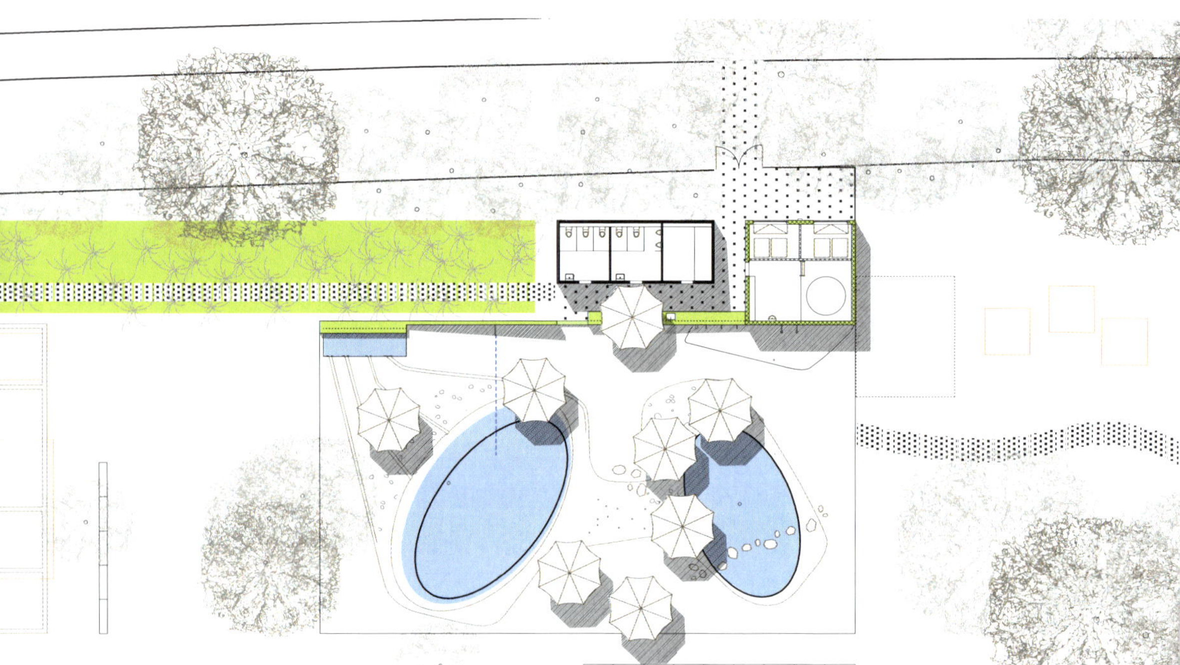

a curtain from a light green wall block at the southwest corner, pouring onto an inclined plane. Children can try to stop the waterfall, and even if the laws of physics win out in the end, the tots have had a lot of fun.

This elemental game is formally articulated in two depressions which the architects interpret akin to the imprint of wet swimming trunks on a terry cloth towel.

As it is so easy to become engrossed in playing here and forget how time is passing, a fleet of parasols, just tall enough for the grown-ups, protects the children from too much sun. In an unusual arrangement, the parasols form an unexpected hovering plane over the visitors' heads, integrating the spatial setting, rather than merely being accessories.

Astounding what you can make out of **a piece of pleated felt** of about 30x80 cm! **Regula Verdet** worked through all the possibilities in **2004.** The simplest one of all: you can place it flat on a piece of furniture as a protective cover, for instance. Oh, really? You can roll it up and use it as a neck support. That's not exactly earthshaking, either. You can fold it lengthwise twice, crosswise to the pleats, fasten it with a pin, and use it as a headband. So far so good, but not particularly exciting.

——— Or you can wear this strip of fabric as a **head-**

covering. It can be a deformed cylinder towering over your head as a cap. It can be adjusted to a certain degree, or squeezed into shape. And then: A certain length of the lower edge is folded up, and it turns into a hat. The opposite is a further possibility: folding the upper edge down. Astonishing things happen. And this is the astonishing part: Folding it by hand overcomes the stiffening effect of the pleats. Once you stop applying pressure with your hand, the ribs take over again and stabilize the newly-shaped form immediately.

The **Sunniberg Bridge** is the most spectacular section of the new highway in the upper Prättigau Valley. The alignment that was accepted in 1989 after a lengthy planning phase features several expensive tunnels, including the 4530-m-long Gotschna Tunnel, to bypass the town of Klosters. Approximately one kilometer before Klosters, the highway crosses the river Landquart at a height of about 60 m. An extraordinary and scenically well-integrated viaduct was to support the great effort for careful landscape protection by means of the costly tunnels. Therefore, in **1994**, the Graubünden Department of Transportation commissioned three engineering firms in the canton to carry out studies with this aim. **Christian Menn**, professor of bridge design at the Swiss Federal Institute of Technology Zurich and an authority recognized worldwide, was invited to serve on the jury to select the winning proposal. None of the submitted studies, however, convinced the jury in all respects. Menn writes to us concerning the further course of events: "Years earlier, I had already studied the idea of a **cable-stayed bridge** on tall piers and had built a model of one that was exhibited for several months at the Department of Civil Engineering of the Swiss Federal Institute of Technology. After we had judged the submitted designs, out of sheer curiosity I asked myself what I would have proposed, and when I saw that model on my bookshelf I scrutinized it as to its suitability for the Sunniberg Bridge. It quickly became clear to me that, using optimal spans with regard to scale, topography, geology and watercourses, a bridge of this kind could be placed very effectively in the valley's physical setting, that the prescribed alignment of

the highway offered certain advantages in terms of structural design, and that the bridge could be given great transparency in the landscape. I then showed the model and a schematic sketch to the architect Andrea Deplazes, also a member of the jury, who immediately recognized the design's potential and strongly recommended it to the head engineers of the Department of Transportation for execution, in spite of the delicate situation concerning the awarding of the contract. Had Deplazes not intervened with such conviction, my proposal certainly would not have been pursued. ——— It was clear that only a firm that had been involved in the prior studies of the project should be commissioned with the further design of my conceptual idea, and due to its reasonable bid, Bänziger AG in Chur was chosen. ——— The procedure by which the proposal was selected and the contract for the design awarded was very unconventional, and a political controversy was possible. Therefore, in order to avoid further complications, I was no longer invited to project management meetings, but only rarely to site inspections. So the story of how the Sunniberg Bridge came to be was not an entirely happy one. ——— Construction of the bridge began in 1996 and was largely completed by 1998. The viaduct served as access road to the tunnel construction site; most of the excavated material was transported across the bridge. The official inauguration took place only in 2005. The parties responsible for the project were: Project management: Canton Graubünden Department of Transportation;
 Architectural consulting: A. Deplazes, Chur. Conceptual design: Christian Menn, Chur. Design and technical

supervision: Bänziger AG, Chur." ——— The following description of the viaduct is also based significantly on C. Menn's writing: The Sunniberg Bridge, 525 m long, with a 500-m radius of curvature, crosses the Landquart River about one kilometer below Klosters at a height of approximately 60 m above the river. Resting on four piers, the cable-stayed bridge traverses the valley in five spans, the middle three of which measure 128, 140, and 134 m. The usable width of the bridge deck is 9 m, with one traffic lane in each direction. ——— The goal of the conceptual design was to optimally integrate the structural system in its surroundings, taking into account above all the proportional relationships between the span distances and the cross-sectional dimensions of the piers, pylons and bridge deck, as well as the topographical, geotechnical and river engineering factors of the site. The structural system itself was to be transparent and have a uniform cross-sectional typology. It was also to make visible its interacting spatial behavior and the flow of its forces. ——— The conceptual design has only two guiding architectural principles: one, the use of the shortest possible (15 to 17 m high) pylons on the tall piers and, two, an extremely slender roadway deck in order to avoid the effect of a barrier across the valley. All other aims mentioned above were fulfilled purely through the optimization of structural measures, which, without resort to architectural means, led to the extraordinary form, the expressive structure of which is achieved through unadorned structural honesty. ——— The entire bridge system is longitudinally fixed by jointless connections of the deck at the abutments, which is possible be-

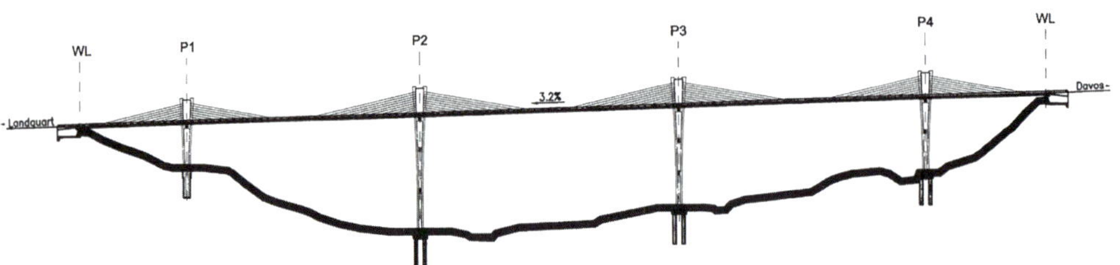

208/209 Before the official opening of the bridge, 2005

Photographs during construction:
the cantilever construction method

Side elevation and profile of the valley

The viaduct after opening

cause of the deck's horizontal curve. Temperature-induced changes in the length of the deck are compensated by lateral shifting of the system – that is, through changes in the curvature. For this reason, the bridge requires neither joints nor bearings, a considerable advantage with regard to maintenance. ——— The curvature of the bridge necessitated the outward lean of the pylons, as vertical pylons would have caused the cables to overhang the bridge deck, compromising the required vehicle clearance. The piers pick up the angle of the leaning pylons and carry it downward, with the tallest piers gradually assuming a vertical profile. The harp-like arrangement of the cables minimizes distraction caused by the intersection of cables in the motorist's field of vision. ——— The deck was built in 6-m-long sections using the cantilever construction method. Unusual forces – eccentric loads caused by strong gusts of wind, for example – had to be taken into account during construction. ——— Virtually the only visible portion of the expensive bypass of Klosters, the bridge is seen as a new area landmark. The cost of the viaduct exceeded the most economical alternative by approximately 15%, or two million Swiss francs in nominal terms – a small sum compared to the several hundred million required for the comprehensive landscape protection measures undertaken. Already, a number of other bridge projects have adopted Sunniberg's groundbreaking concept. The Sunniberg Bridge stands as an exemplary work in which high aesthetic demands and complex technical considerations are brought together in inspiring harmony. It is recognized internationally as a masterpiece of engineering.

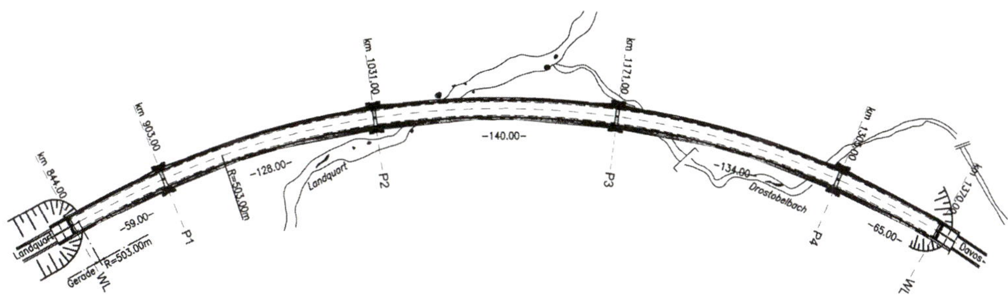

Bridge, seen from the right-hand side of the valley

Alignment of the road, with the tunnel entrance on the right

The **Worb-Dorf railroad station** (Berne Region) serves several functions at once: as the terminus for passengers of the Berne – Solothurn regional railway line, the RBS; on the upper level as a park & ride **parking deck** for cars, and at night as a **depot** for the rolling stock. In 2003 the architecture firm **smarch** – Beat Mathys & Ursula Stücheli was commissioned to realize its design in cooperation with the engineers **Conzett Bronzini, Gartmann** (cf. p. 230) because its proposal to unify the required program in a single structure was the most convincing competition entry. ——— Paradoxically, it is the least public of its functions, that of nighttime depot, that most strongly defines the architectural character of the building. It is transformed into a depot when a 16 m wide rolling grille at the front and two gates at the back close off the platform area and the tracks. The design's main feature is the side walls of tensioned metal bands which protect the station from intruders along its 130 m length. These bands are 23 cm wide and 1.5 mm thick, are made of duplex steel (chromium steel with a low coefficient of expansion) and are woven in and out around the round columns. Held together in pairs by a double-bolted clamping ring, they cling to the columns by their tension alone. Since the clamping bolts are mounted only in every other panel and, in their vertical succession, each band is offset from the next, the surface appears like shimmering woven fabric. The columns at either end, drop-shaped in section, take up the entire pulling force exerted by the bands. ——— In an inviting gesture to the village, a canopy projects generously from the head of the building, the seeming counterpart in a plug-and-socket connection to urban Berne at the other end of the railway line.

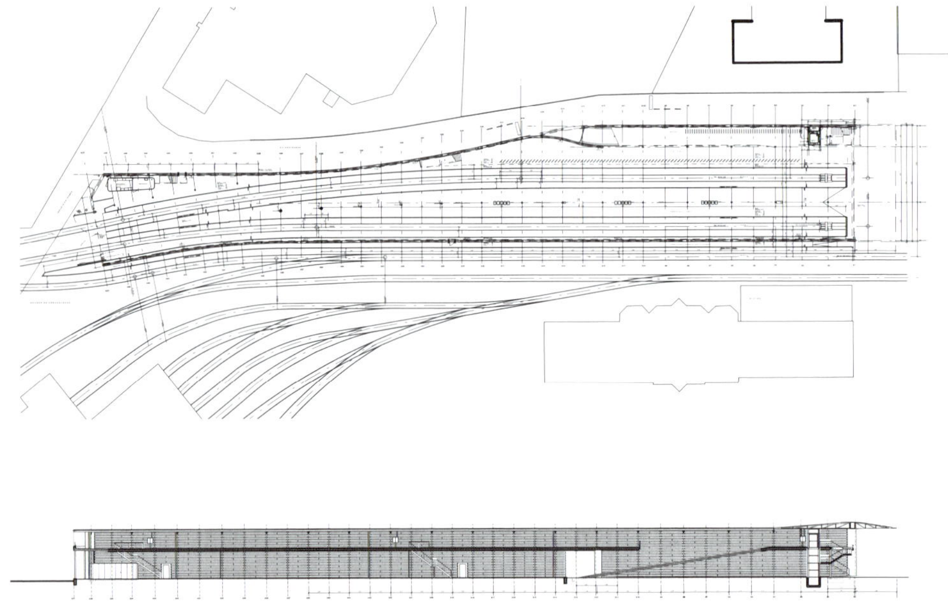

Floor plan and longitudinal section with the ramp to the parking deck

A marble track is the perfect example of an object with a predictable course. A marble follows the simple law of its sloping path. Small children enjoy it immensely, feeling vindicated when the object does what they have commanded it to do. Then, however, comes the moment in a child's life when predictability begins to bore and he or she can penetrate the mystery. Wilhelm Kienzle (see p. 48), for example, recognized this problem and gave his track the form of a sounding campanile: A marble is sent on a journey through the interior of a wooden tower, falling on replaceable aluminum bars and playing a corresponding melody. ———

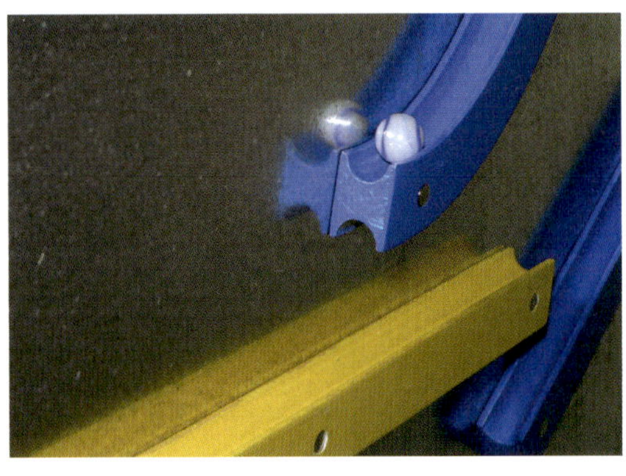

Samuel Jäggi had a different idea in **2005**. His **magnetic marble track "Kugelspill"** boasts tracks which are detached sections; modules (1 straight, 1 curved element) whose magnetic back stick to any metal surface – a radiator, a door, a refrigerator – where gaps are allowed and there is no need for a continuous track, where the marble can drop over a void, where angles can be changed and boundaries can be tested. How fast can the marble go, how slowly can it roll, what accelerates it, what slows it down? How much speed is needed to perform a loop-the-loop? The elements of colorfully varnished wood are always available for experimentation.

The main idea behind this colorful piece of furniture is the search for the reduction of form until only lines and planes remain. Numerous designers have taken on this challenge in the past decades, and it is current once again. Designers **Markus Huber Recabarren** and **Michel Nigg** created their **"Plattenspiel"** shortly after graduating in **Zurich** (designed in **2000**, now sold by www.moobel.ch). Wilhelm Kienzle's "Embru" shelving unit without screws (1931) was one of their points of reference. Whereas Kienzle's unit is constructed of L-shaped beveled sheet metal supports fitted into grooved boards, in this case contemporary manufacturing techniques such as computer aided milling extend a *series* of possibilities until they become a *field* and, even further, a *space of possibilities*. Thanks to the precision-cut grooves, "Plattenspiel" can be put together with no need of screws, racks, or other fittings. It is a three-dimensional structure made of MDF, aluminum and Plexiglas. The weight is carried by aluminum sides that are set at an angle and keep one another in place, the necessary bracing is also achieved in this manner. By simply shifting the colored planes sideways in the grooves, volumes can be created or dissolved. The unrestricted arrangement of the sliding colored planes allows for play on many levels. The translucency of the colored planes, through which some of the contents are visible, is an element which beautifully individualizes this piece. The object dissolves the spatial limits of the individual compartments and refers to the world of images of modern architecture and concrete art.

The idea of a subterranean extension lent itself well to the expansion of the main building of the **University of Zurich** with a large auditorium. In **2003**, it was developed from a terrace element of Karl Moser's significant late-Jugendstil building (1914), which overlooks the old city of Zurich. For their design, the architects **Gigon / Guyer** (Annette Gigon and Mike Guyer) and the artist **Adrian Schiess** referred strongly to the geometry of Moser's building, but were able to work structurally with longer spans. Therefore the lecture hall was predominantly developed in its width – three axes wide. It is accessible from Moser's transverse hall, painted light green and white and situated at the base of the spectacular grand atrium. The transverse hall had been divided into a number of seminar rooms and had attracted little attention for a long time. Two stairways lead down from the hall between wooden surfaces painted berry red to the auditorium below, where almost 500 seats are arranged in long tiered rows. The front projection wall and the lecturer's platform (at the lowest level) lie below a horizontal skylight, from which the access of daylight can be regulated by blinds. The ceiling is well articu-

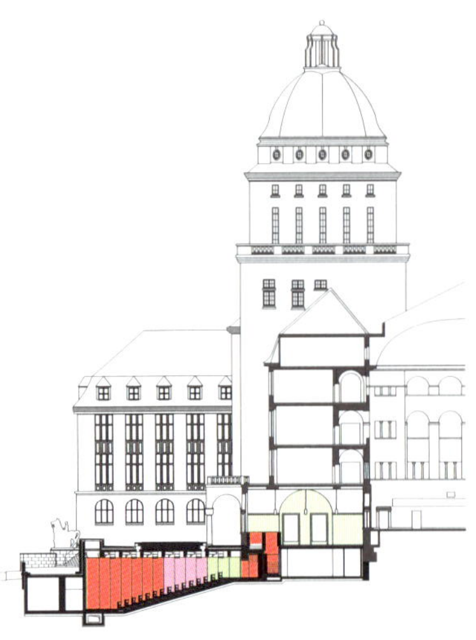

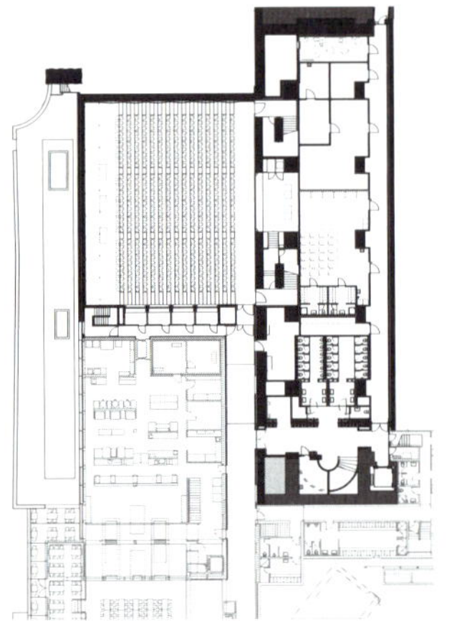

Longitudinal section of
the auditorium and the Moser building

Floor plan

Terrace facing the city and new basin

lated by means of a repetition of box-like headlights, which relates to the seating arrangement of the tiered floor. Above the auditorium, a new, shallow rectangular water basin serves as a mirror rippled by the wind. ——— The main features of the design were simply and logically taken from the existing surroundings. An important point is how the lecture hall was built in details. The acoustics played an essential role. Rectangular rooms with parallel walls are acoustically unfavorable. Nevertheless, the architects found ways to provide impeccable acoustics in this rectangular lecture hall. The auditorium's rear wall is sound-absorbent, as are the desk front panels, and the right-hand side wall is almost indiscernibly zigzagged. These moves prevent flutter echo and additionally produce fascinating reflections caused by the ceiling lights. The most amazing color effects were conceived by the architects and their artist friend Adrian Schiess to interact with the elements. The faded and horizontally layered color of the elevation facing the city and the same washed out dusty pink of the basin above the auditorium is a significant contrasting element to the rectilinear simple building volume.

Interior showing the color effects

220/221 The large auditorium

Dancer **Ania Losinger** and instrument builder **Hamper von Niederhäusern** jointly invented the **floor xylophone Xala**. Originally a flamenco dancer who danced to live music, Ania Losinger sought a new form of expression. Since **1998**, she has been making her own music, as rhythmic as it is melodic, in spatial choreography with the Xala whose name is derived from the Basque percussion instrument Txalaparta. She plays original compositions, the work of others, and, seldom, improvisations. She uses flamenco shoes to play the sounding beams, occasionally also utilizing long rods to create additional percussive parts. ——— Xala is a 4.5 square meter sounding parcel of land as it were, graceful, and shimmering in red. Above it lies the space in which the dancer re-creates the connection (ancient, but lost in the course of humankind's development) between dancing and making music. Technically speaking, Xala is about harmonizing the mechanics of movement of the standing, turning, or even reclining artist (for instance: turning around one's own vertical axis, or a circular leg movement) with the laws of acoustics (the length of the sounding beam, that is, the position of the artist, determining the pitch, and the character of the sound dependent on the type of touch). Above all, the artist must fulfill this ideal by performing it for the audience not just as a brilliant technical feat, but as an artistic (visual and acoustic) achievement. ——— In technical terms, Xala is an ingenious structure that consists of 6 rectangular metal frames, each equipped with 4 sounding beams ranging between 1.2 and 1.4 m in length and made of east African padouk wood. All the beams have the same width, their differing depths determine the pitch. Xala weighs 400 kg. In 2004, it was complemented by Xala II, which differs from its predecessor in the design of the sounding beams and the more sophisticated tunability. The most difficult technical challenge was designing the bearings of the sounding beams so that the artist's weight would not suffocate the sound. The solution was to design each sounding beam's points of support so that the transverse oscillation specific to the particular note also passes through them. Both instruments are played solo as well as in a variety of musical formations.

Alongside her work at the "Eclat" agency, **Renate Jaberg** ran her own **"Graphic Salon"** for smaller private commissions mainly from the Zurich club scene. With subtle irony, the name hinted that she used a ladylike cultivated form of expression to advertise her salon, rather than the heavy-handedness commonly found in the design community. In **1998,** Phillip Meier, the proprietor of the **"Substrat"** lounge, commissioned her to design its **monthly program flyer.** Since his specific concept involved allowing the DJs and VJs free rein in designing the evening, Jaberg also worked with the motto **"carte blanche in a black box".** The postcard-sized flyers, printed double-sided, were its visual interpretation.

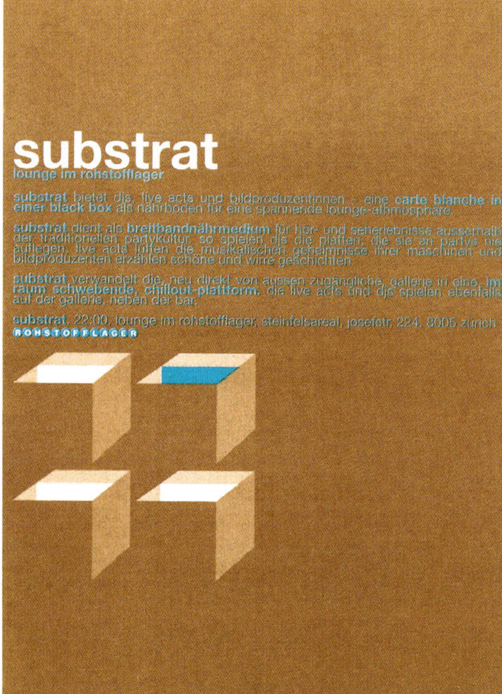

Flyer, June 1998, front on the left, back on the right

Renate Jaberg worked according to these flexible guidelines for more than two years, up until the lounge's premises were torn down, producing a series of postcard flyers which were characterized by the parallel perspective, Helvetica font, her use of two colors, and the motif of a cube with a card inserted. Over time, the flyers developed into a series of variations with a fairly broad color spectrum and occasionally recurring color combinations. ——— The "Graphic Salon" also designed card series with other motifs and modified forms, yet always remaining faithful to a stated design logic, for "Granulat", "Stratos", and "Taifun – Rote Fabrik", other Zurich clubs.

Flyer, November 1998, front

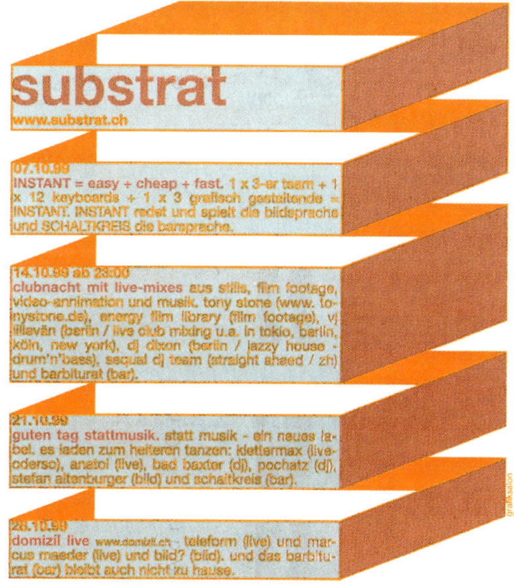

Flyer, October 1999, front

What is design? We answer this question differently today than in the era of functionalism, when it was a question of meeting basic needs. In today's affluent societies, or, more accurately, for those who belong to them, our needs are more than met. Instead, we have developed an appetite for ideas. And then? Design means searching, but how can we prevent that what we have found from seeming "sought"? When design is only the execution of the already-known, the step toward boredom or formalism, or both at the same time, is quickly taken. ——— The **Collenberg/Ponicanova** team (Patricia Collenberg and Zuzana Ponicanova) show us an entertaining way out of the dilemma between the difficulty of meeting elementary needs and the boredom of surplus. They are not seeking out the form to fit the function, they also define the function. Like an equation with many unknowns, many variables must be made to fit together. ——— In this case, they not only designed a **short jacket** in **2002**. Instead, they ripped into the task of creating something new and different. Their jacket is a jacket, but not only that: It is a **set** of three halves for the wearer to choose from. There is a back and two fronts, with no clearly defined outside and inside. Parts which can be combined in different ways, buttoned together in myriad combinations Additional sewn-on elements can be pulled through a slit and be made visible or – take it or !eave it! – remain concealed. Depending on your mood. Outside to outside or outside to inside or inside to inside – like this, and like that, or like this. Only a twosome can have ideas like this. Collenberg/Ponicanova give the simultaneous search for function and form a fitting, funny, fabulous expression.

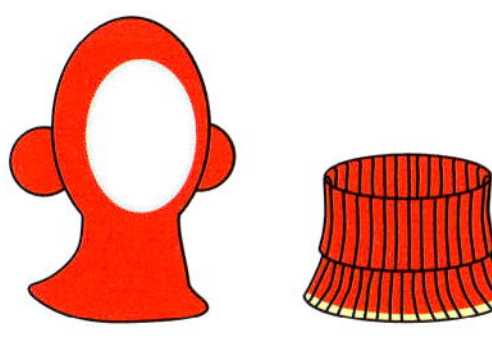

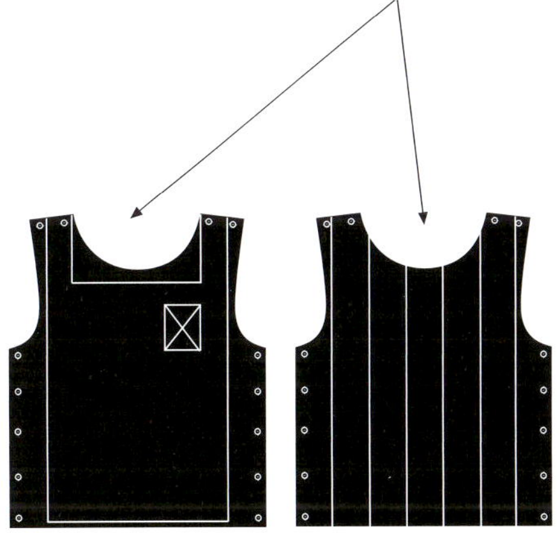
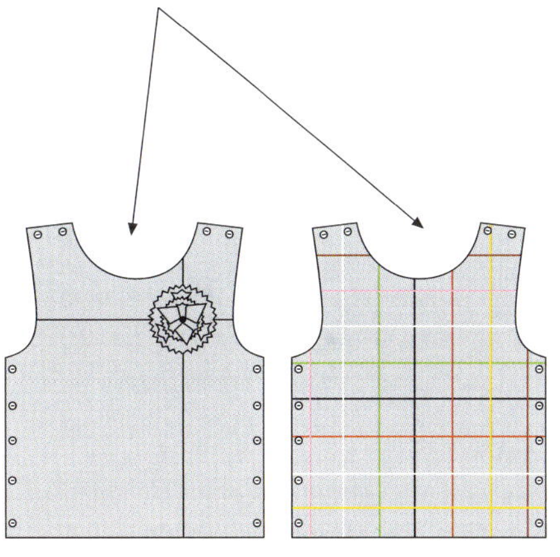
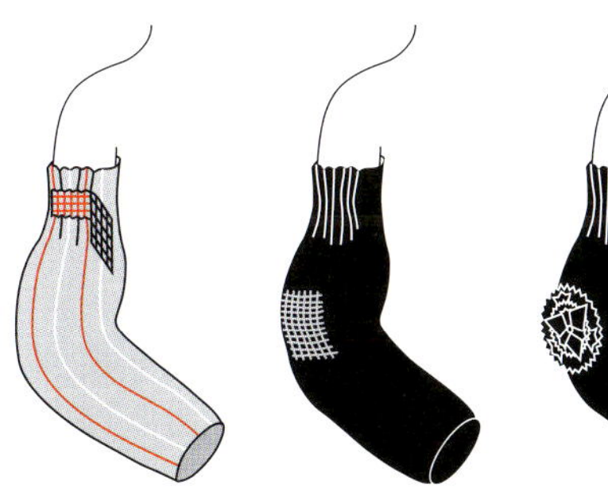
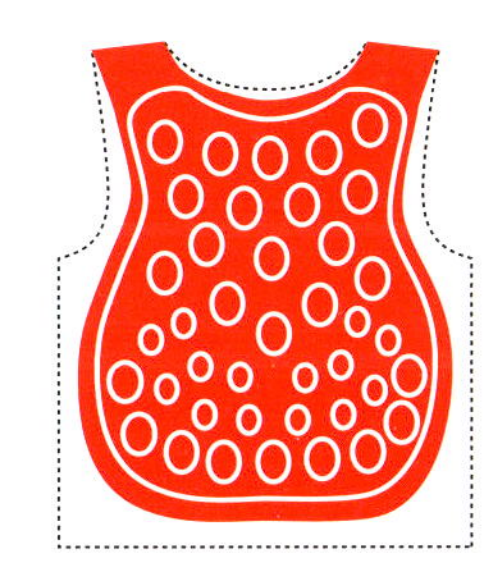
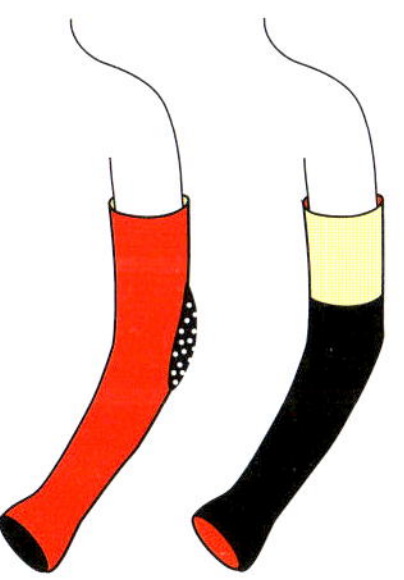

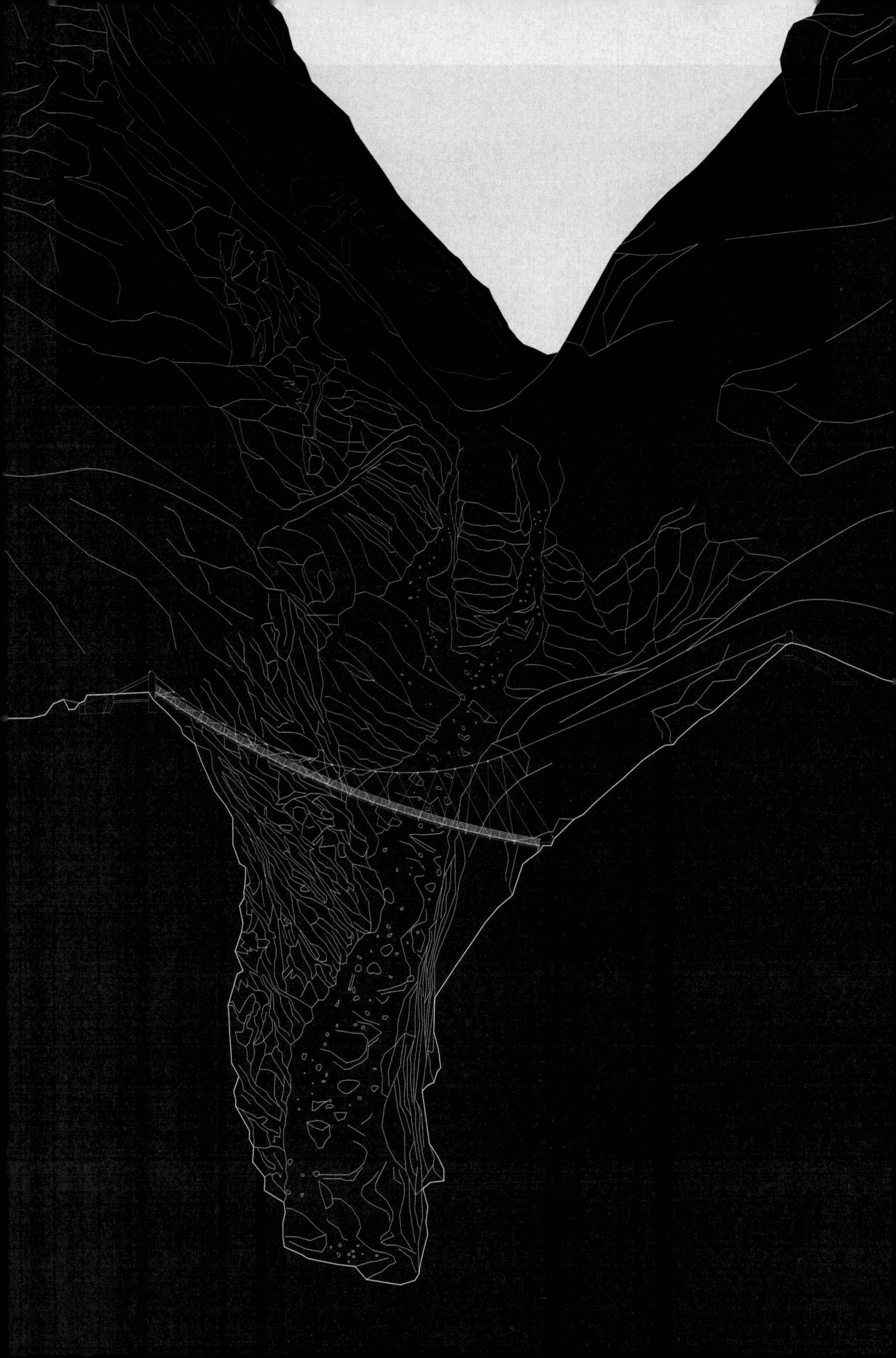

Suransun's **pedestrian bridge**, spanning a side arm of the Rhine Valley, was built in **2005** to replace the lightweight, prefabricated triple-chord wood truss bridge, also designed by the engineering firm of **Conzett, Bronzini, Gartmann** (Chur), which had been flown in by helicopter in 1996 but was unfortunately destroyed in a rock-slide in 1999. The new footbridge was built not far from the first, at a broader but safer crossing over the gorge. The client was once again **"Kulturraum Viamala"**, a private organization founded to both run an 'eco-museum' and preserve the cultural heritage of this dramatic landscape. In this case, an old mountain trail was to be revitalized. At first, it was questionable whether it would be 230
possible to build a footbridge able to bring hikers safely across the drop of almost 70 meters. The engineers came up with the solution of "hanging stairs" borne by a prestressed hybrid suspension/truss system that gains 22 meters in elevation over a distance of 56 meters across the deep ravine. Prestressing was necessary in order to reduce oscillation and make safe crossing possible. This was done by building the stairway as an inverted arch and placing it between the walls of the gorge, thus creating a preload force on the cables under tension. The main cable is attached asymmetrically, the much higher southern mooring ensuring a relatively low tensile force at the abutments. The system of prestressed suspension cables rising in two vertical planes from the flanks of the bridge distributes the tension on the main cables evenly. ——— Ten individual glue-laminated arches connected side by side – two under the stairway and four to each side of it – make up the inverted arch structure that carries the stairway. The wooden arch construction rests on crossbeams of galvanized steel (at 3.6 meter intervals). The broad base of the supporting structure both lends stability in the face of horizontal forces (wind) and also acts as a blind, providing psychological support to hikers wary of heights. To prevent hikers from stumbling, the rise-to-run ratio of the individual steps is adapted to the progressive gradient curve. Safety was a high priority in the design – to achieve it, an "invention" was necessary.

228 Cross-section through the gorge, view of the footbridge

229 View from the south

View of the staircase and the hybrid suspension/truss system

This **house** on the **Rigi-Scheidegg** was built in **2003** by two couples – including designing architects **Gabrielle Haechler** and **Andreas Fuhrimann** – for their own use. A concrete basement story serves as the foundation for the two-story prefabricated wooden house, whose elements (ceilings and walls) work together structurally as a whole. The elements were transported to the construction site by helicopter and assembled in a single day. ——— With its irregular hexagonal footprint, the house has neither distinct front, back and sides in the accustomed sense nor a conventional roof. None of the exterior walls meet at right angles and the roof planes slope down to the perimeter in several directions. This irregular "faceting" gives the house an appearance suggestive of a great boulder deposited randomly in these foothills of the Alps, while the wood cladding alternatively makes the impression of an axe-hewn block. Only the interior walls on the upper floor are laid out in rectilinear order. However, since these do not touch the outer walls at any point, it is the polygonal course of the exterior walls that decisively shapes the space. The organization of both stories

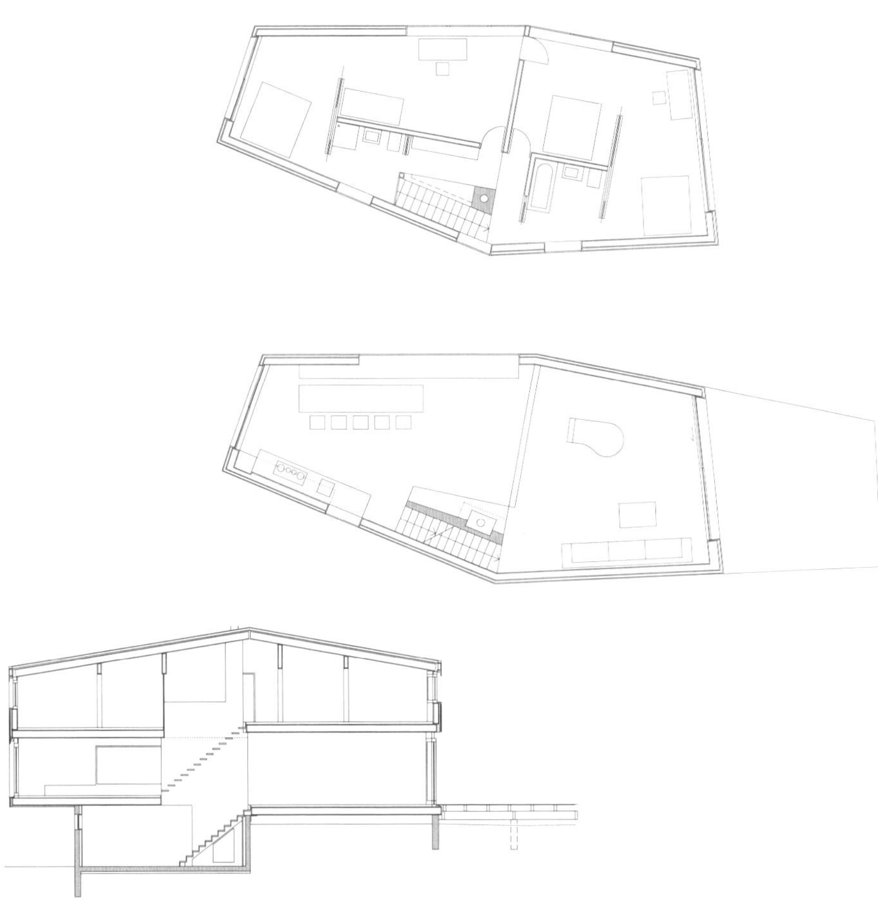

Upper level

Main level

Longitudinal section

defines a circular path around the interior – explicit in the spacious shared living area on the ground floor, more implicit on the upper story, which is divided into two private areas. The stair and the layout of the rooms lend a counterclockwise "spin" to the interior circulation. This effect is heightened by the variously shaped windows, which are akin to pictures on the walls; rather than frame static views, however, the windows enable the occupants to experience a *passing* land-scape as they move through the house. ________ The house makes simple use of low-cost materials. The use of the eye-catching knothole pattern in the interior plywood paneling is inspired in part by contemporary art. The siding of overlapping planks is often seen in utilitarian farm buildings. Despite its freshness and air of independence, in some of the themes it addresses this house is a tribute by the architects to a distinguished colleague of two generations ago: Hans Leuzinger, who, between World Wars I and II, found a language for modern alpine architecture in his designs for a number of houses for the Swiss Alpine Club.

RELATIONS

Steven Holl on Switzerland, the Guhl chair, the Swiss residence in Washington D.C. and on the relations of forms. ——— The feeling I had when I first visited Switzerland in 1970 was a very inspiring view of charcoal-stone-mountain faces in contrast with white ice and snow. So the "snow on the alps" inspiration was a very deep-seated memory for me, a connection to the Switzerland that I have so often visited over the years. ——— We developed this materiality in charcoal stained rough board form concrete, and contrasted it structural glass plank which is almost ice-like in opaque white (it is a semi-gloss materiality). ——— This slate grey color is also related to the slate grey shingles in this Washington D.C. neighborhood. ——— The design was the unanimous winner in an anonymous competition. All ten competing teams had a strict list of requirements and program from the Swiss government. Our concept which revolves around a diagonal line of spaces from courtyard to interior – to alignment. The distant view of Washington Monument was central to the idea. The "precinct" walls which mark out the rectangle of the building in the green landscape make it like an urban place. In this little urban courtyard inserted the cruciform volume of the main building as well as the caretaker's quarters. It was important spatially that the "precinct walls" line up with the end walls of the main building allowing outdoor spaces and indoor spaces a real interrelation. I like seeing the Willy Guhl chair very much against this background. ——— While the 1950's were great for modern furniture design, only a few pieces really became classics, such as Arne Jacobsen's Ant Chair and the Guhl chair. I first saw the Guhl chair in Zurich around 1980 and thought then that it was a real classic design which should be more prevalently used. ——— When you say "Playfully Rigid", and I see this image, I think also of contrast, light and heavy, curved and straight. ——— I believe if you have an image of our Turbulence House from Abique New Mexico which is all curvi-linear and you put a Donald Judd chair in front of it, you essentially have the same relation as with the Guhl chair in front of The Swiss Residence.

Excerpts from a conversation
with Steven Holl, by Sabine v. Fischer

IMPRECISION IS IMPORTANT

“If you want that people should do something, you have to avoid precision. A very precise work is very difficult. If you admit and tolerate imprecision, the thing will be more feasible for people. There is so much evidence. It makes obvious that building techniques that rely on precision demonstrate that such an incredible precision makes no sense. I don’t mind it, but if you want that people really participate – then it makes no sense. This is my personal belief; I believe in things, which are not expensive, and in things, which are not precise, and in things, which everybody can handle. That is certainly a personal attitude. And I believe that the word luxury could have a meaning, but this comes more with the content of the things, than with the material quality. You can draw a nice thing simply on the sand with your finger, so it is not the material that is important. This is a very old-fashioned view. ——— […] I knew Jean Prouvé, when I first came to Paris. We thought we could do a project together. Then, we could not do it because he went broke; he lost his factory, in 1957. So we stopped the project. It is this cylindrical project. I designed it and for the technical realisation, I was looking up Prouvé, and we got along very well. And then, he thought that we could produce the elements for it in his factory. But the project got stopped when he lost the factory, and he joined this wagon factory, which designed the métro and so. ——— […] When I think of Prouvé, I think he was really a craftsman, because he really liked simplicity. One of the things was that at the same time he was very much into precision, this was a difference. Like Konrad Wachsmann (we were friends) – but Wachsmann as well was a maniac for precision! But, later, he said to me: “You were right.” ——— […] Because I was not looking for it – he was the one who was looking for technical perfection. I was looking for imprecision, for usefulness, and for the possibility that somebody would make it the wrong way. That is very important. You cannot count on that everything will be done well. So for popular utilization it is extremely important that it can be made the wrong way. ——— You know, Wachsmanns and Prouvés generation were completely influenced by the industry, and by Bucky Fuller, completely. So my generation was less trusting. Frei Otto is from my generation: He works with enormous precision but at the same time with techniques, which allow for a certain imprecision. ——— […] These cylindrical units, that I told you about before, we could not sell them because they were too cheap. All commercial firms were not interested, because the have no profit from it. That is characteristic. Individual persons were interested, but we could not produce it out, because, as you know, to put up a production line, that is another thing. ———
I think the ready-made of Duchamp was a great idea. The things is that he stopped with the ready-made, with the object, that he did not go further, into accumulation, into constructing with these objects, but by any means, it was a revolutionary thing: that anybody could do it.”

Excerpts from a conversation
with Yona Friedman, by Sabine v. Fischer

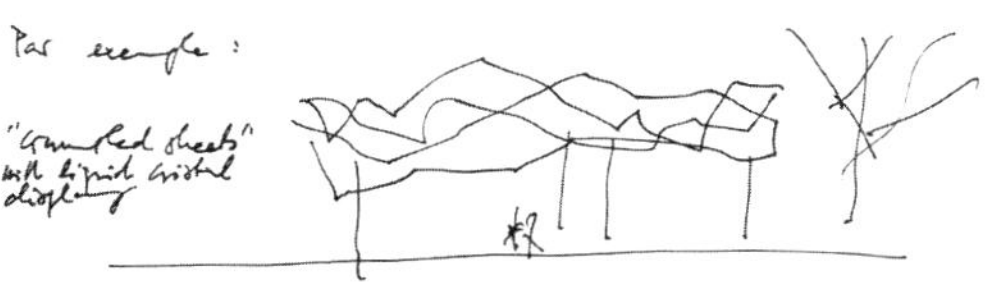

Roof for a Paris school made from crumpled metal mesh, 2003

Since the establishment of the **Fotomuseum Winterthur** in **1993** (it was renamed **Fotozentrum Winterthur** when it was consolidated with the Swiss Foundation for Photography in 2003) the four graphic designers **Käti Durrer**, **Hanna Koller**, **Jean Robert** and **Trix Wetter** alternate in designing its posters and invitation cards. Robert and Durrer work collaboratively as a twosome, Koller and Wetter work on an individual basis. ______ The Fotozentrum devotes its activities to all forms of photography since the invention thereof. The technical medium remains essentially constant, the societal circumstances and the artistic aims of an exhibition cover an extensive range. ______ Thus the graphic designers chose to adhere to an overall design concept, presented as a fixed set of rules to the pub-

Four invitation cards from a good ten year span
(with identical posters, except for the missing opening hours)

lic, yet remained internally flexible, based on the following four maxims: **1.** The upper half of the format is reserved for a photographic subject and typographical elements are positioned on the lower half. **2.** If the photographic subject has a horizontal format, then it shall also be placed as such on the poster; the typography is then turned at a 90° angle and placed vertically. **3.** If the subject is in an actual portrait format, it is then turned 90° counter-clockwise; the typography is placed horizontally. **4.** The background colors of the lower area of the format, the choice, size, and color of the font, may be freely determined within the above listed parameters. ——— This set of rules has remained in place and has proven itself well suited to bestowing the institution with a clear public profile.

Jürg Fontana's **lighting rail** from **2000** is an acrylic light diffuser set in a stainless steel plate folded into a “U” and illuminated from behind. The light source is a fluorescent tube. This lighting rail distinguishes itself from typical lighting strips by a second main characteristic apart from providing light: the acrylic insert is designed so that slides can easily be inserted and exchanged – best of course are pictures you've taken yourself! A sliding magnifying glass allows you to look at any slide you want more closely. The long model (154.5 cm) has space for thirty slides, the short one (100 cm) for twenty. Will the lighting rail be placed over the breakfast table, or next to the bed? Will it lie on the floor? Lean against the wall? It can be anyplace where the eye can behold it. This is how easy a design concept can be, how simple design can be. And everyone is happy, because the good old slide – Aficionados say: a high-resolution medium! – has not yet lost its place in the world.

Bookbinder and designer **Christian Guggenbühl** designed another interpretation of the **corrugated cardboard chair** in **1996**, and his firm Format Guggenbühl has been carrying it ever since. The two planes glued to one another are cut, perforated, and fluted in such a way as to permit folding them and tying them up with a ribbon to a three-dimensional structure. The ingenious interlocking of wedges achieved by folding the cardboard permits the astoundingly sturdy chair to support even a weighty person. Not only does the chair combine advantageous static properties with being very lightweight, it also permits the person sitting on the chair to place his or her feet underneath its front edge. This is a comfortable, resilient and formally convincing chair. It can be folded and unfolded several times without losing much of its stability. The ribbon is not just an element to tie up the package when transporting the chair, but serves to tension it when it is assembled, thus contributing to its charm of appearing provisional.

In a series of steps begun in 2000, **Novartis Campus** – the pharmaceutical company's research and development campus in Basel – is being developed according to a master plan by architect Vittorio Lampugnani. (Novartis is the result of the 1996 fusion of Ciba-Geigy and Sandoz.) The site, floor space, and function of the Forum-3 building were all stipulated in the master plan. The architecture firm Diener & Diener won the international competition in 2001 for the realization of the project. Vienna architect Gerold Wiederlin and artist Helmut Federle collaborated with Diener & Diener on the proposal. The users of the building, which was completed in 2005, are employees in Novartis' pharmaceutical Research and Development department. ——— Occupying a site at the outer border of the campus at Voltaplatz, the building is accessible only from within the campus. Despite the building's exclusivity, the friendly, enigmatic expression of its facade manages to avoid snubbing the city. 240

241 The front facade facing the campus and
plan of the glass panes in the front facade (black = space)

All five exquisitely designed and furnished stories are entirely glass-enclosed and feature a second "skin" of variously colored panes on all four sides. These are not "windows", they have no physical function such as climate control, but instead condition the building's psychological climate. Like glass feathers, they appear to flutter about the clearly delineated functional framework. The 1,200 glass panes are mostly muted in color. In 25 different forms and 21 commercially available colors, they envelop the prismatic building in a minutely worked out composition that apparently follows no preconceived rules, yet abides by a deliberate order. The panes obscure the regular structure of the vertical supporting rods to which they are attached. More intensely colored panes, fewer in number, are gathered in the areas of the building's vertical circulation, along the division between floors, and around secondary rooms. The architects have nonetheless avoided applying any obvious hierar-

chy in their placement. This is not public art on a building ("Kunst am Bau"), but rather a building as art. ——— The ground floor, set back deeply from the face of the building, at once serves as prelude, transition, and access to the three stories above. These open-plan office floors are interspersed with small glass-enclosed, snail shell-shaped meeting rooms. Far from chaining them to isolated desks, the architecture clearly calls on the employees working in these offices to interact and communicate with one another. A loggia running the perimeter of the building behind its glass facade offers an open space for both relaxation and temporary work. ——— Lavishly worked in wood, a freestanding winding stair – reminiscent of Salvisberg's architectural design for Hoffmann/la Roche in Basel in the 1930s – rises from the first to the fourth story. So too does the exotic garden at the building's southwest end, designed by landscape architect **Günther Vogt**.

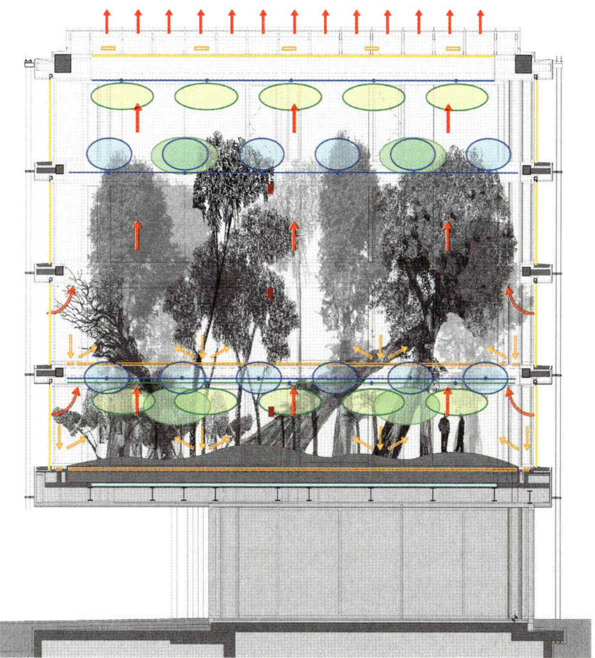

Lounge (to the south)

Section of the exotic garden

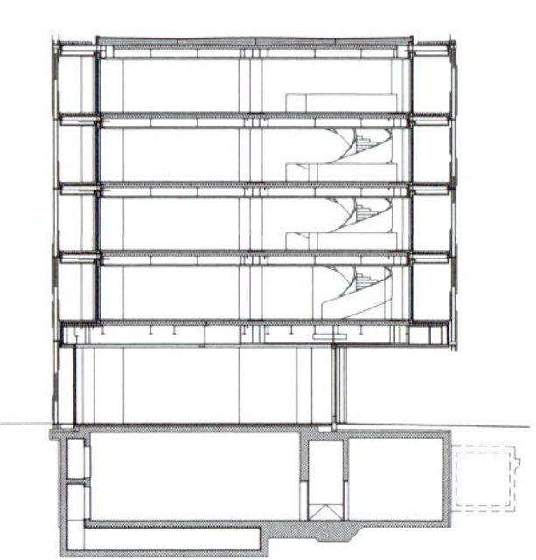

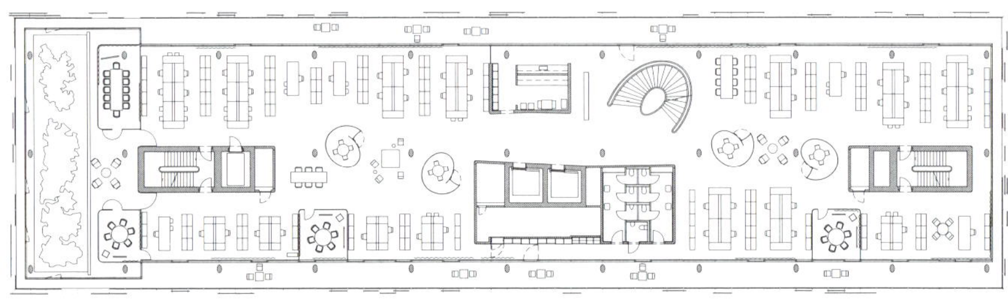

Office floor with small meeting room

Section

Floor plan, second floor

The wooden connecting staircase

Just a few kilometers away from Neuenburg, the **Gorge d'Areuse** (Areuse Gorge) cuts into the Creux du Vent. In 1999, the architecture firm **Geninasca Delefortrie** was commissioned with a new **footbridge** at the scenographically important location. It was realized in **2002.** ——— The design responds to the different characters of the two banks in an impressive, almost poetic manner: The north bank is steep and narrow, the south bank wider and more open. The footbridge reflects this difference by creating a sickle-shaped space that accentuates the transition from one bank to the other. The footbridge widens when one crosses it from the narrow side, and it narrows when one crosses from the more open side. Pedestrians

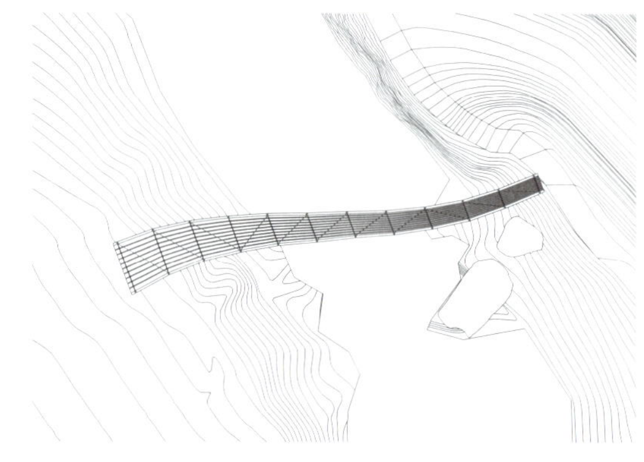

crossing this doubly curved bridge walk on ground Jurassic marl held by sheets of metal. Narrow wooden slats make up the side walls and the "ceiling". They turn the footbridge into a fascinating cage-like space with irresistible suction due to the horizontal and vertical curves. With its width and height identical (3.5 m), the southern end forms a square. The northern end is just as high, but only one-third as wide (1.15 m), shrunk to a narrow rectangle. This structure represents an idea of form, not an engineering concept. This design approach was feasible given the span of just 27.5 m. The supporting structure is a framework of hollow structural steel painted brown and barely discernible from the wooden slats.

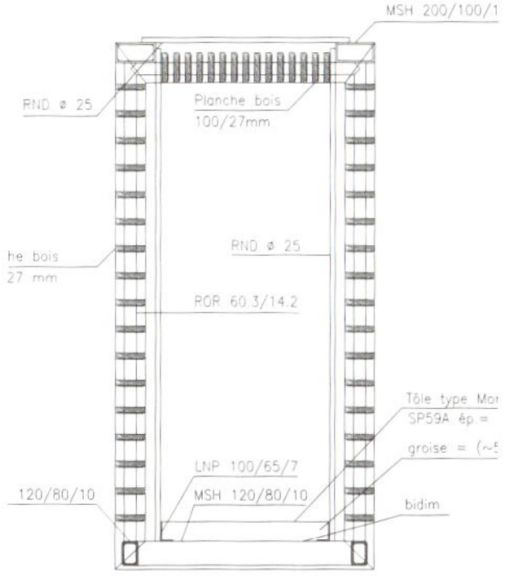

For **Antoine Cahen** and **Claude Frossard**, the two founders of **Les Ateliers du Nord** (ADN), several aspects of design are important which usually do not get the attention they deserve in the design discourse. Even in today's highly computerized living and working surroundings, they seek a product design idiom that demands more of the users than simply pressing a button; the use of a product should include a gestural element. The fact that children reputedly love to use the Nespresso coffeemaker satisfies these intentions. ADN seeks to achieve "transparency" in appearance – in the sense of the word that the form of the appliance should express something about its functionality. The studio aims to bring forth designs meant to accompany the user through life for a period of some time, not just ephemeral accessories for an imagined lifestyle. If someone says that an ADN design *somehow* reminds them of this or that animal, the designers welcome such comments in this implicit form, even if direct affinity in form is never the goal. And by the way, one of ADN's first works around 1984 was the development of the first mouse for personal computers, in collaboration with the firm Logitec (While it is true that the "mouse" had been invented years before in the US, it could not be used in the existing computer centers and had therefore come before its time).

——— The **flashlights** for Leclanché are a subtle and technically sophisticated design. The

batteries are coated with colorful plastic, and the head containing the light bulb fits directly on top of them. The light is turned on by twisting the ring-shaped lamp socket. ——— ADN's most recent work in **2006** is the new hydrant for the **von Roll** foundry. The commission was to develop a **hydrant** to be used internationally that could be installed by a single worker (previously, two workers had been needed). The design permits a reduction in the hydrant's weight of 20 kg and consists in separating two parts in the constructive design: a straight, simply cut off pipe and a top made of cast aluminum. This upper piece can be mounted in a variety of ways and can be adapted by adding additional components (depending on the existing infrastructure in the particular country). The *ordinary* hydrant, *by the way,* is a surprising example of market liberalization. As a capital good, it had evaded consumer preferences, and for decades it was the expression of a de facto monopoly in Switzerland and an immutable standard. Deregulation created competition and launched an international battle for market shares. Von Roll would like to compete Europe-wide and even globally with this new model. And this: For ADN, the hydrant, too, is an important element in one's living surroundings. Even if you pass it by day in and day out, seemingly heedlessly, and *by the way,* it still makes its mark in your individual world.

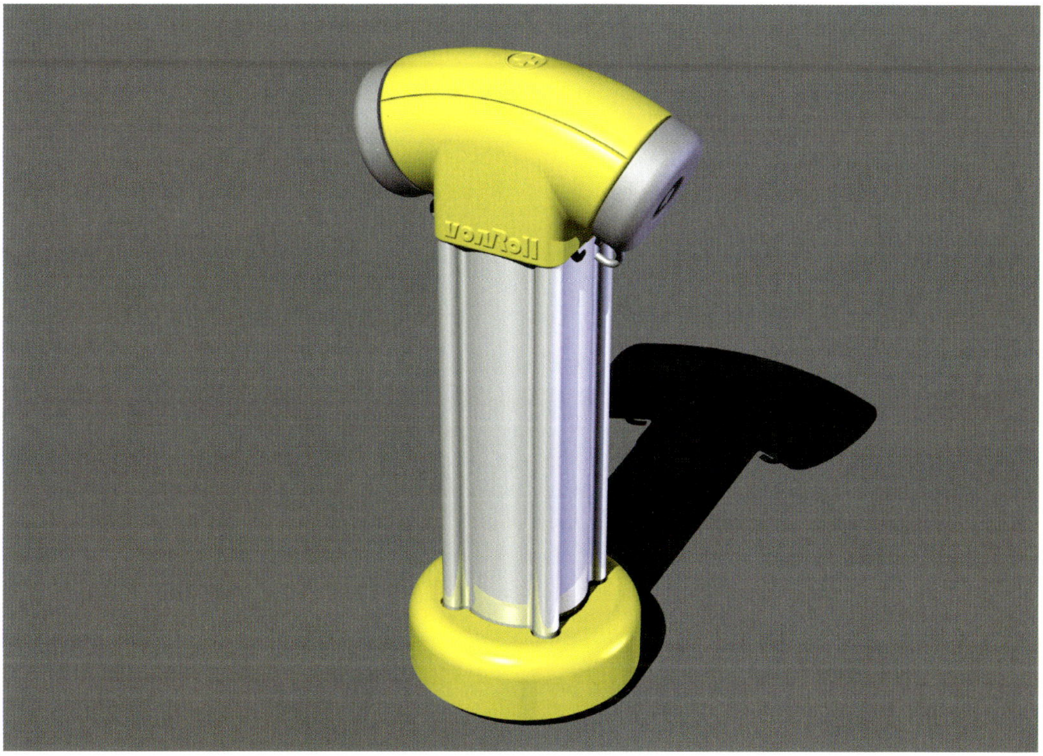

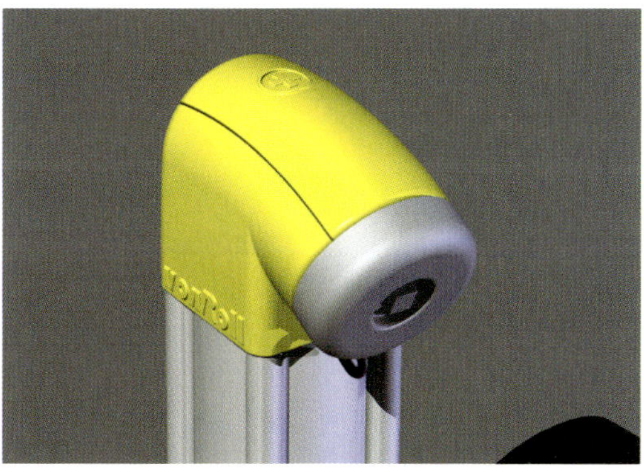

Norm's two members, **Dimitri Bruni** and **Manuel Krebs**, have earned a reputation as creators of all sorts of projects. These are productions that at times are humorous and informal, but they go even further. Among other techniques, Norm has developed its own typography software and has applied it in famous public locations, such as the Cologne-Bonn airport in 2002. ______ One example of their contemplative working technique is their **book** about the actor **Bruce Lee**, published in **2005**. Years ago in Beirut, they discovered a cheap paperback about the kung-fu star, his biography in Arabic, containing poor quality photographs, obviously second or even third generation reproductions of the originals with Moiré effects, cloudy patches, badly cropped edges and the like (Bruce Lee – The King of Kung Fu. His life, his art, his films and his death, from "New Sports Series" originally published by The Modern Library, Beirut 1975). They reproduced the book page for page and printed it in the exact size

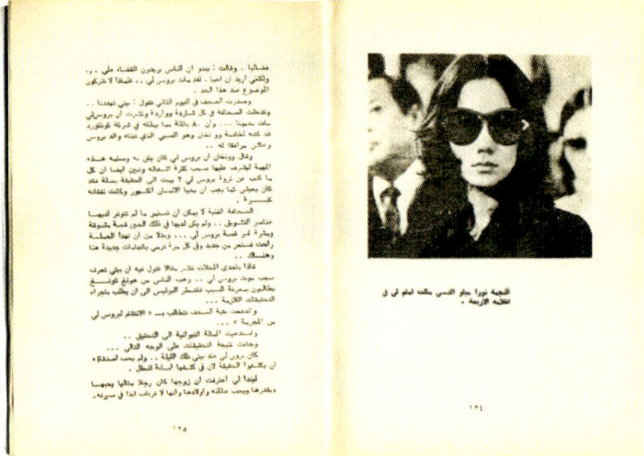

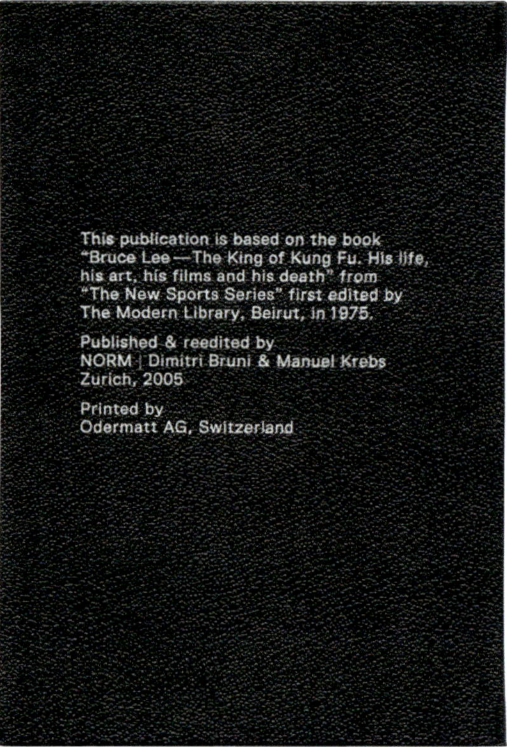

and page order with one critical difference, however – they removed all the script printed information. The illustrations are in exactly the same positions as in the original, but printed on glossy paper using a high-quality printing process – which also produces a facsimile of the original version's inferior printing quality. The narrow hardcover book is bound in familiar, pitted black imitation leather. The front cover bears the title "Bruce Lee" in embossed silver lettering and the back cover notes the bibliographical data and the Norm address as the only printed information in the entire book. Turning the book into a piece of fine art provides an official disclaimer of its original trashiness. It s difficult to describe the fascinating effect this book has after its entire printed wording has been removed. The only thing that remains as a subtext is the peculiar language of the images in a book, which only speaks to us visually, without words.

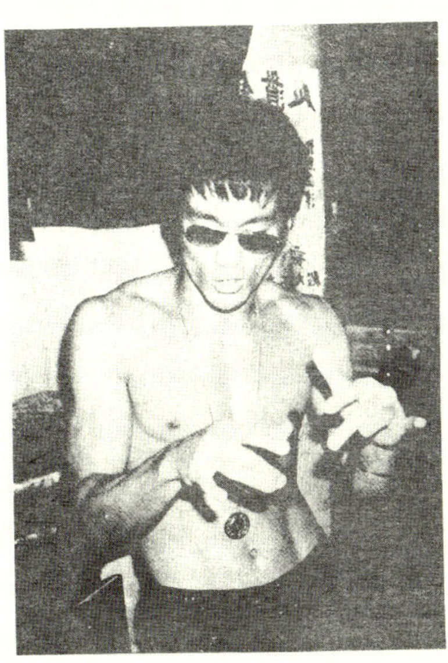

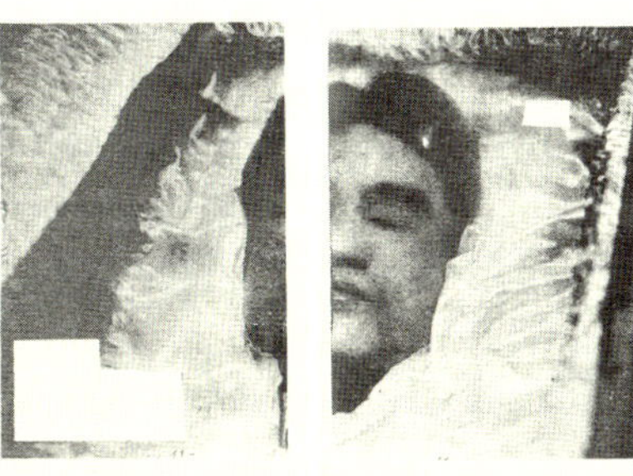

The **Mattenhof School** in **Zurich** is an extension of a school designed in the 1950s. Both the form and material of the new buildings reflect the character of the pre-existing structures. However, that which in 1955 was the usual signature of the epoch became in **2003** a subtle commentary upon the same by **B.E.R.G. Architects**, Zurich (Sibylle Bucher, Christoph Elsener, Michel Rappaport, with Volker Lubnow). The entire complex has a uniform appearance, and is nevertheless granted a fresh accent by a new building and a covered outdoor area to be used during breaks. Motifs such as skylights and externally mounted, projecting window frames, as well as the roughly-textured surfaces, are given a contemporary interpretation. The school was redesigned for all-day use, a prototypical innovation in Zurich at the time. This led to changes in the character of the individual classrooms, which are no longer just places of earning, but places of living as well. ——— The main entrance is on the middle floor (on schoolyard level). A central circulation

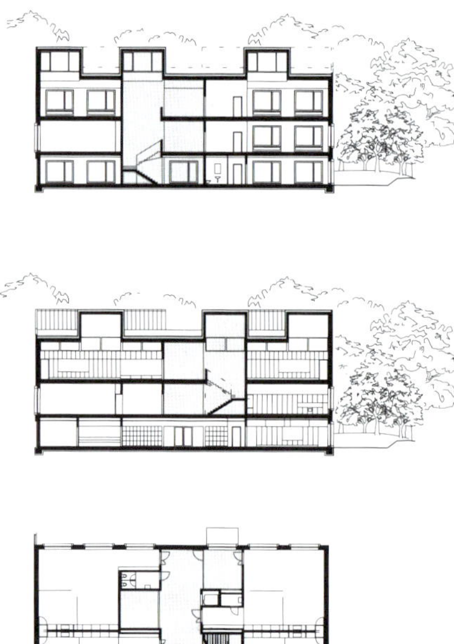

Stairways

Longitudinal sections of the two staggered staircases

Floor plan of the second floor

View from the schoolyard

Classroom

Facade detail

space with curving stairways and large anterooms dominates all three stories of the building. The staggered staircases are more than a means of "vertical circulation"; they describe a slanting path through the three-story building. Classrooms incorporate the area of both small-group rooms and cloakroom and are therefore especially generous in feel. Connecting doors between adjacent rooms enable two classes to be combined for specific teaching units. The layout of the seemingly spacious rooms also allows for group work and team-teaching. On the top floor, chimney-like skylights cut through the slightly pitched ceiling creating a bright, shifting "ceiling landscape" in the individual classrooms and throughout the building. Ever-changing daylight moves through the classrooms and stairwells throughout the course of the day. ——— When children draw houses, the chimney is often set crookedly on the roof and the windows are larger than in reality. Here, architectural reality comes astonishingly close to the world of children.

A leisurely summer evening by the lake. Brought a bag with something to eat and a newspaper… The picnic is over, the food's eaten up, the paper's been read and burned in the bonfire. The bag is empty. Now it's ready for a different use. Zippers open up, fabric is turned inside out, edges are reconnected in different ways. A new alliance forms. The shoulder belt becomes the sleeves, the bag transforms itself into something else, a light jacket for the journey home. **Susann Brütsch** presented this study on the theme "Transformation" in **2003** and found a fitting name for it: **Yaksak** – "from the jacket to the sack to the jacket" (the J in "Jacke" is pronounced like a Y in German). One design criterion was a minimum of surplus material – no doubling, no leftovers. You take along only what you really need.

S

O

255 STAND
DER
DINGE

NEUSTES
WOHNEN
IN
ZÜRICH

N

W

P

F

11. BIS 22. FEBRUAR 2002

EWZ UNTERWERK SELNAU
SELNAUSTRASSE 25, ZÜRICH

VERNISSAGE
MONTAG 11. FEBRUAR 2002
18.00 UHR

TAGLICH VERANSTALTUNGEN
FÜHRUNGEN
WOHNDEBATTEN
XENIX FILMPROGRAMM
BAR, LOUNGE
AUTORENTAG

ÖFFNUNGSZEITEN
TAGLICH 17 BIS 23 UHR
SAMSTAG 14 BIS 04 UHR
SONNTAG 10 BIS 17 UHR

WWW.STANDDERDINGE.CH

Amt für Hochbauten der Stadt Zürich

ETH

I

Tania Prill, poster “Stand der Dinge”, Zurich 2002

Tania Prill and **Alberto Vieceli** have been running a joint studio for visual communication since 2001. Tania Prill designed the poster **"Stand der Dinge"** ("The State of Affairs") for the **2002 Zurich** Good Building Award. They are seldom commissioned for posters, however. The Prill–Vieceli studio focuses on book design for numerous publishers. Their **books** all have very different characters; like other designers working today, Prill and Vieceli have no personal style that can be recognized at a glance (this is true of their posters as well). Their books differ in format, in typographic treatment, in their use of images, in weight–in everything really. What they have in common is the premise that each and every book project has its particular individual right to recognition. When it appears, who writes it, what content is associated with it, its desired audience, the goals to be reached by publishing the book, the publishing line it is to appear

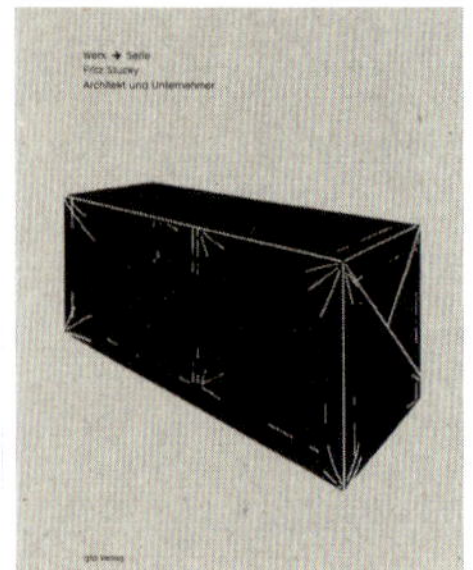

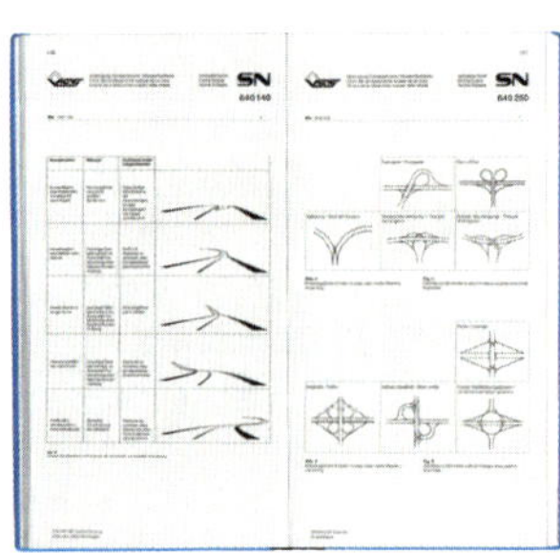

Book *Fritz Stucky*, 2006

Book *Autobahn*, 1999

in – the list of criteria is long. They could simply be answered and checked off one after the other. Publishing houses routinely work this way. Not so Tania Prill and Alberto Vieceli (and other important designers of their generation). The subject matter gets first say, or, more precisely perhaps, the potential of the subject matter. What is the direction of its gravitational pull? The design process consists mainly of the endeavor to resolve this question. One can only find the answers in the process of gradually settling into the work. Answers which transcend purely visual criteria. How does the book feel, how does it lie in the hand? What sound do the pages make when you turn them? This is a different level of attentiveness, its prerequisite is a resistance to technical overkill. Making the right choice instead of just accumulating means to an end – this is the task they have taken upon themselves.

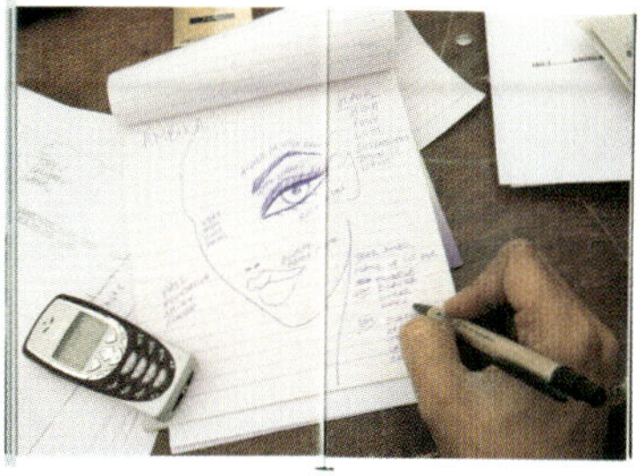

Book *Khadi. Textiles of India*, 2002

ELLA

8 - 10 dA
NovembER O2
12 aVELS
ratUrA
Dis LittE
da
M 1317
A DomaT

Since 1991, the **Rhaeto-Romanic Literature Festival** (Dis da litteratura) take place every November in **Domat-Ems**. Since **2002**, graphic designer **Theres Jörger** and photographer Susanne Stauss have put a visual face on the three-day event, which has a new thematic focus every year. Themes in recent years were, e.g.: Surrendering tradition – telling stories, Literature for children and young adults; in 2006, the focus will be on female figures and women authors in literature. The two designers collaborate closely, conceptually interweaving their contributions.

They are fascinated with letters, which are more than the written and printed symbol of speech. The designers place them as "letterally" materialized words in space and bring them back to the surface through the use of photography. Their work is a translation which repeatedly jumps from the front to the back: from the intended meaning (wording) – the setup of space with its corporal letters – to its photographic representation and thus from the graphic design and the choice of the means of design (format, type of fold, type of paper, color) to its transposition into the medium of print. For the 12th Literature Festival program, for example, they used rubber stamp letters; not the tool-face itself (the rubber), however, but rather only the side on which the letter is printed for "operative control". Complicated? No, not at all. But in this way, and by the use of blurred depth of field, the image tells of being an image which was made and not just a reproduction. It follows that literature, language in general, is something to be talked about.

Atelier Oï in **La Neuveville** is a team boasting a wide variety of skills and interests cleverly brought together. It takes on commissions from small product packaging to the design of public spaces, from living accessories to exhibition design. The three founders, **Aurel Aebi**, **Armand Louis**, and **Patrick Reymond** (and additional partners today), have been working together since 1991. Their projects have an attitude which mediates effectively between "Swiss German" functionalism and the francophone desire to strike a pose. With the partly bilingual origins of its partners, and its geographical location near the language border, Oï is a living example of the "idée Suisse". The name Oï – derived from l'oeil, eye – is a witty neologism and stands for a state of mind set on relishing work and life (and not just the posture of pretended ease so often encountered). ——— The two works chosen here convey this well. The **valet stand**, which Oï designed for IKEA around **1998**, is a Plexiglas surface with two circular holes at the top and bottom. The vertical plane is subdivided by wave-like cuts running crosswise. When the valet is hung on the wall

by the top hole and pulled apart lengthwise, the waves splay outwards. When the panel is attached at the bottom as well as at the top, stabilizing the "deformation", slits and compartments emerge to hold small objects: scarves, gloves, hats, lightweight jackets, etc. The design is not only about a technical procedure in the interest of function, it is also about the user's enjoyment of understanding this procedure. Design as a form of communication – it can be more than just a slogan.

The bed for the Wogg collection **(Model Wogg 24)** has a frame made of extruded molding, its legs are the juncture points. A belt reminiscent of those used by moving companies runs around the base of the bed holding the tension, and also bracing the legs. Sliding rails set in the framework allow the use of the space under the bed for a variety of containers which can hold bedclothes, books, clothing, etc. The mosquito net accessory, modeled on the protective covering for a fruit pie, might seem like a gag, but it can unexpectedly be quite useful, provided the mosquitoes are on the outside.

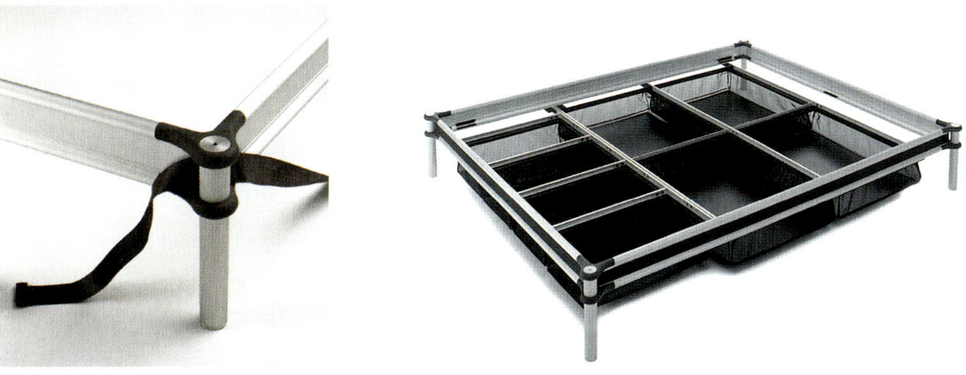

Where Lake Constance again flows into the Rhine, in the village of **Eschenz** (near Stein am Rhein), stands a straw bale house built in the fall of **2005** by architect **Felix Jerusalem** for a young family. Because the site is in an area of archeological importance (Roman Rhine crossing) and is barely above groundwater level, the house is set on concrete piles, with a cellar dug under only one small area. A monopitch roof rises along the longitudinal axis of the house, reaching a height of two stories over the living room. The construction material is particularly unusual: compressed straw panels for the floors, walls and roof. It is an environmentally attractive, surprising and light material with good structural and thermal characteristics, and is also quick and easy to handle (sectoral planning by Hermann Blumer, Herisau). Since there is no manufacturer for the panels in Switzerland as yet, these panels were imported from Mecklenburg. The construction took only three months, the installation of the 25-cm-thick straw panels only a week. Floor plan and section correlate with one another closely. Both are organized around a concrete core containing the wine cellar and, on the ground floor, the sanitary unit. 262

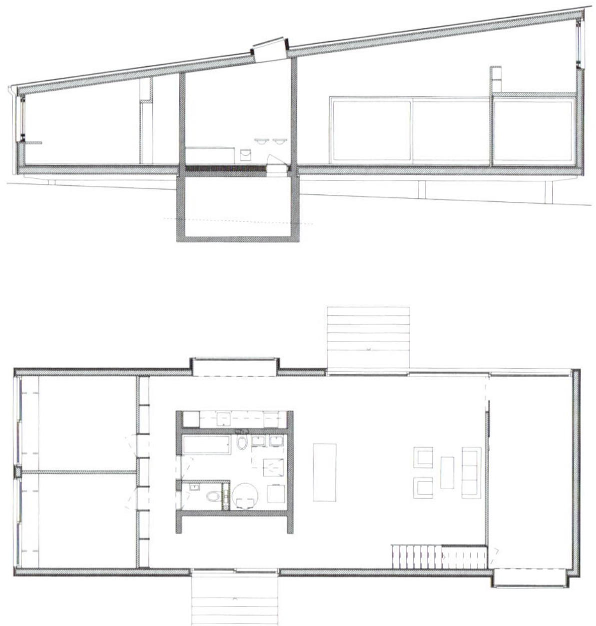

Two children's rooms are located at the lower end of the house, while the parents' area with a study loft occupies the higher end. Large sliding windows open onto the garden; in the kitchen, a box-type window punches through the exterior wall. Greenish translucent reinforced polyester corrugated panels (Scobalit) protect the straw panels, allowing a view of both the straw panels and the framework to which they are fixed and lending a sense of depth to the facade. ——— Despite its low cost of construction, the house also makes contribution to installation art that is worthy of mention. This is the product of a collaboration between the architect and Lausanne artist Karim Noureldin. Noureldin's intervention consists of three rings of different sizes painted on the floor. Depending on visibility and the viewer's angle of vision, they either have the quality of accidental discoveries or make strong reference to the house's floor plan. They impose themselves on the house's functional zones, «disturbing» or corroborating them, and also alludes to the historical importance of the building site. On a golden child's ring recently excavated from this site, the word Intius, from which Eschenz is derived, is engraved.

"If something is used in contradiction to its inherent traits, another name must be found for it." This sentence and drawing by **Yves Netzhammer** (on a **poster** for a **1999 exhibition** at the **Museum Allerheiligen** in **Schaffhausen**) could easily be smuggled into a first grader's primer. A good teacher with an interest in philosophy could probably turn it into an enjoyable lesson on applied epistemology (in other words, the practice of the theory of knowledge). ——— At first glance, Yves Netzhammer's proximity to surrealism is evident. His approach is, however, much more systematic and thoughtful. He is the morphologist at the mixing console for figures, who has one image emerge from another, and yet another out of that new image. The means of transformation is drawing, which is always based on linearity, even when it is modeled almost three-dimensionally with virtuosity. The contours of these figures are sharply defined and their surface appears to represent a non-porous separation between themselves and the vacuum surrounding them. Netzhammer's world consists of solids; the meaning of his world is the exact opposite. The two sides of an invitation to the **Helmhaus Zurich (1999)**, are an example of how he imagines images – in this case a "dead" object. The beam of the flashlight, which we know to be cone-shaped, is a true geometric cone. And the surprising reversal of color accentuation on the back of the card turns the flashlight itself into a figure of light. This is an example of bipolar graphic design; there are long chains of images in Netzhammer's oeuvre, analogous "inventions" which fan out and break their ties to one another. He reveals formal and semantic processes of transformation within the logic of a world of images which is both personal and societal. In mechanics, a transmission is a contraption whose different parts are connected compulsorily, so that movement of one part is transmitted to another part. Working more mysteriously and not with cogs, Netzhammer's works are akin to an image transmission. The power of the images – their danger, the abyss they show us, and also their surprising truth – is transmitted from him, the author of the images, to us, the observers, at a junction, sometimes found within one image, sometimes between images.

Invitation card Helmhaus Zurich, 1999, front and back

Yves Netzhammer, exhibition poster Museum Allerheiligen Schaffhausen, 1999

The story of the Freitags is well-known (the young brothers **Daniel and Markus Freitag**, who lived near an arterial road in Zurich; the view of the heavy trucks struggling across the Hard Bridge toward the Gotthard Pass; the inspiration to make a (1!) shoulder bag out of the plastic tarpaulin of one of these trucks as a personal gift... That was in **1993**. It was crucial that the **shoulder bag** show a detail of the lettering, so that in the leap in scale from truck to accessory, from whole to fragment a story is told about the transcontinental trips the material has taken. The bag's enthusiastic reception instantly made it clear: that's cool! And the inspiration turned into a business idea). ———— In **2006**, the **Freitag Shop Zürich** was built next to the arterial road using 17 stacked freight containers held in place by tension members, joined internally, and decked out with entirely new interiors. The architects were **Annette Spillmann** and **Harald Echsle**. Is this Freitag-Shop symptomatic? Today's idea of the city is no longer the 19th century ideal of the "City Beautiful", neither is it the 20th-century practice of companies seeking a good retail location as defined by this ideal. Following the scenographic thinking of the 21st century, a location turns into a retail location when the proprietor calls the shots in the trend-setting scenes. That is the case here, in the "written-off" (amortized) in-between realm at the edge of the central business district that awaits a blast of fresh air like a low-pressure system. The glittering world of the city center couldn't be further from mind. Until now, Freitag lab.ag's sales display consisted of cardboard shelves. As they quickly became unsightly, young designer **Colin Schälli** designed a new sales display in 2006 under a strict budget, made of stackable boxes held together by bungee cord rings. 96 boxes form a unit. It is still undecided whether they are to be made of a biodegradable spray cast material like liquid wood or of recycled plastic. When they come into use, they will congenially bridge the gap in the sequence: product–packaging – sales display – container building.

SHOP
ZH
OOCL
AWS
Occasionen
Mietwagen
Leasing

Graphic artist **Franziska Burkhardt** has designed posters and a variety of print materials for concerts, stores, cultural institutions, and events (Kunsthalle St. Gallen, Helmhaus Zurich, “plug in” Basel, et al.) as well as designing the exterior of the Xenix cinema in Zurich. Her work spans a gamut of genres. In addition to her work at her own her atelier “Buka Grafik”, she collaborated with **Nadine Spengler** until 2004 on various projects under the label “Mäusepolizei” (Mice Police), producing smaller printed works and, in **2001**, a cleverly designed **book** for the **Migros-Museum** which received quite a bit of attention. Nadine Spengler, due to her strong affinity to comics and illustration, is very involved in her own project: the “Pipifax” bookstore in Zurich, which she co-founded and runs with Axel Friedrich. “Let’s Be Friends”, the book created for the Migros-Museum, was a retrospective overview of the five years work of the management team Rein Wolfs and Gianni Jetzer. The two graphic artists’ design concept was innovative: instead of reproductions presenting the exhibitions in chronological order, Franziska Burkhardt and Nadine Spengler created their own alphabetical classification of the museum’s diachronic activities, labeling them by use of a thumb index. For example: A for “A”, B/Building materials, F/Found but not Lost or U/Unknown. The assignment of the artwork to a specific rubric was left up to the graphic designers. This work is difficult to present visually, however it is worthy of special mention here as it exemplifies, in a daring and humorous manner, their approach to design as vehicle of authorship and not just a service. ——— In her other works, Franziska Burkhardt explored myriad paths. She designed the graphic appearance of the “Xenix” cinema (in Zurich) as follows: the envelope of the monthly program is always a 1:1 detail of smaller poster of the corresponding thematic cycle – yet which detail and the positioning of the envelope motif and how it shall appear (mirroring the poster, upside-down, slanted?) is up to the designer commissioned with the poster – someone new each month. These prescribed rules to play by both facilitate and force creative decisions. Here we also show her **poster** for the implementation of the **2003/04** annual cycle of contemporary music **“tonart”** (Bern). ——— Attention to materials and the print medium itself are also key elements of her markedly conceptual work. Such considerations as: How does a certain paper feel, how a program (e.g. for “tonart”) is folded, what happens when it is unfolded, when the cover and back overleaf are fleetingly, yet significantly, momentarily simultaneously visible.

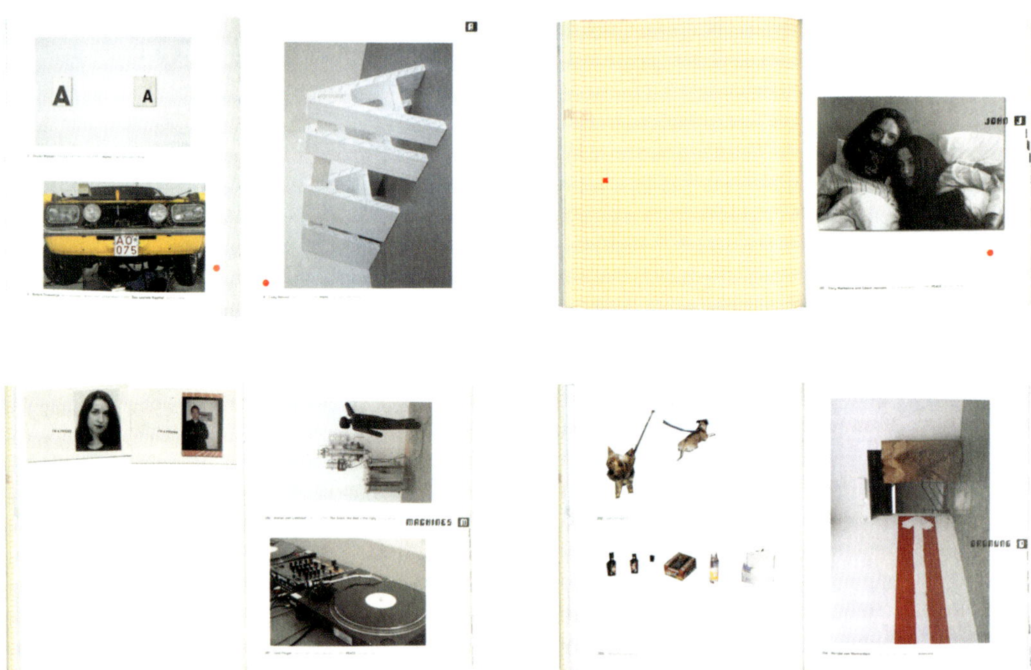

Book *Let’s Be Friends, 5 Jahre Migros-Museum,* Zurich 2001

Franziska Burkhardt, festival poster “Tonart”, Bern, 2003/04

Christian Kerez is an architect and thinker who has not built much yet, but the little he has built is quite substantial. His designs have a high specific density. They emphasize, radically and with fascinating individuality, the basic theme of architecture: the interactive conflict of material and space. For Kerez, it seems that architecture is no longer dualistic; he dissolves antagonisms such as concave – convex, inner – outer, support – load, centrifugal – centripetal, etc. His architectural configurations have a highly intellectual weight without appearing intellectual. This is the crux of the fascination alluded to; first and foremost, the themes Kerez deal with strike the senses and are actually considered as architectural events before the language. See what happens when a corridor is liberated from the rooms it serves, no longer considered a mere conveyer to them, but rather as a directed spatial event. ——— The **Leutschenbach schoolhouse** in North **Zurich** (competition in 2004, currently under construction) promises to become an impressive

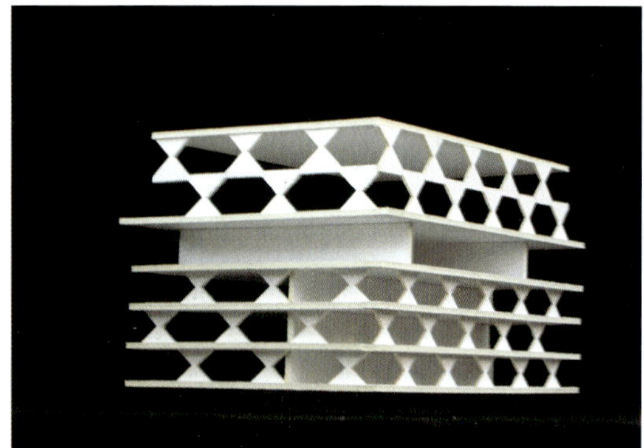

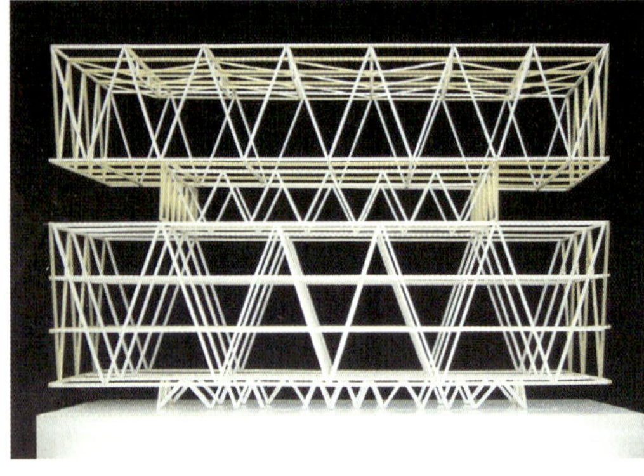

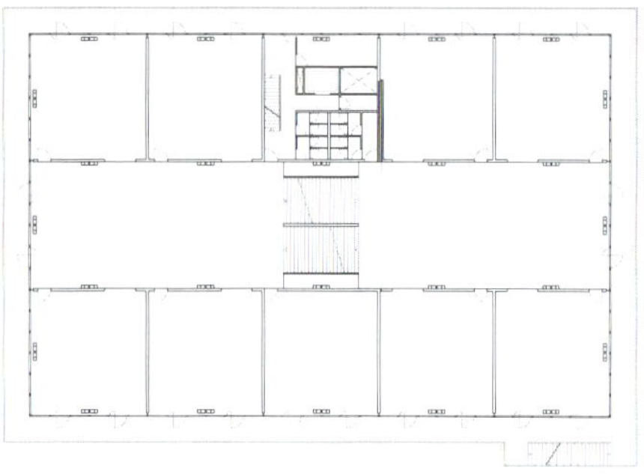

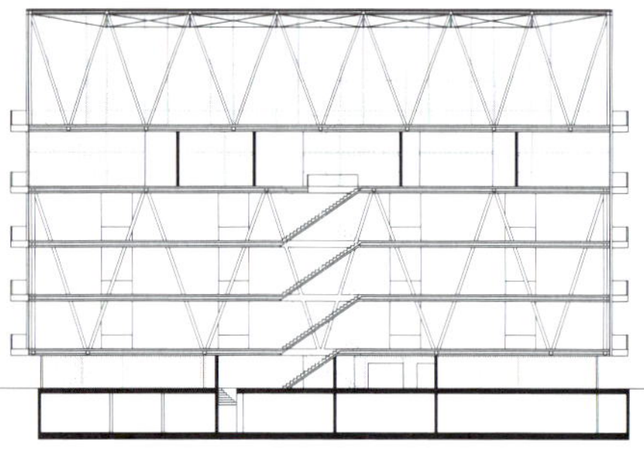

facility in which the functional areas are layered on top of each other to form a large entity. Between the 8 m high double gymnasium at the top and the classroom floors below, there is a separating story that – just like the ground floor – is defined by a horizontal framework, which articulates the individual parts within the whole building. As in other works, Kerez does not consider the supporting structure conventionally as a rising pyramid. The narrow basis of the ground floor is a clear indication of the "Other" which he seeks. The supporting structure of layered, intertwining vertical iron framework binders achieves its effect by means of spatial arrangement in the three dimensions of Euclidean space; and partly also by means of suspended ceilings that give the structure a more active articulation. In their formal treatment, the tensioned parts are not easy to discern from the compressed ones (and are not treated didactically). This gives the glassed framework an even stronger effect. What goes on here? People will wonder.

Two structural models

Rendering at night

Floor plan

Longitudinal section

One of today's topics is the image as an element of reality, more than a mere "representation" (as in the case of usual reproductions) can ever be. **Martin Woodtli** was one of the first graphic designers to use the "Form Z" architecture and design software and attained an unusual level of mastery with it . For his **book "Woodtli" (2001)**, which documents his work, he developed a successful solution to the problem of original and reproduction. In the book, he presents his works in their original sizes, but then recodes the color scheme. For example, a work which is printed in blue and yellow in his book was originally in green and orange. The original color appears on the edge of the page with a corresponding indication of the Pantone shade. Woodtli depicts small posters in such a way that they are also printed in the original size, but with the colors remodulated. Since these posters are significantly larger than the book format, they are then printed as two or three double pages. Two copies of the book opened to the proper pages and laid side by side will then produce the whole poster. Thus, it becomes clear that the term "documented" as used above is not cor-

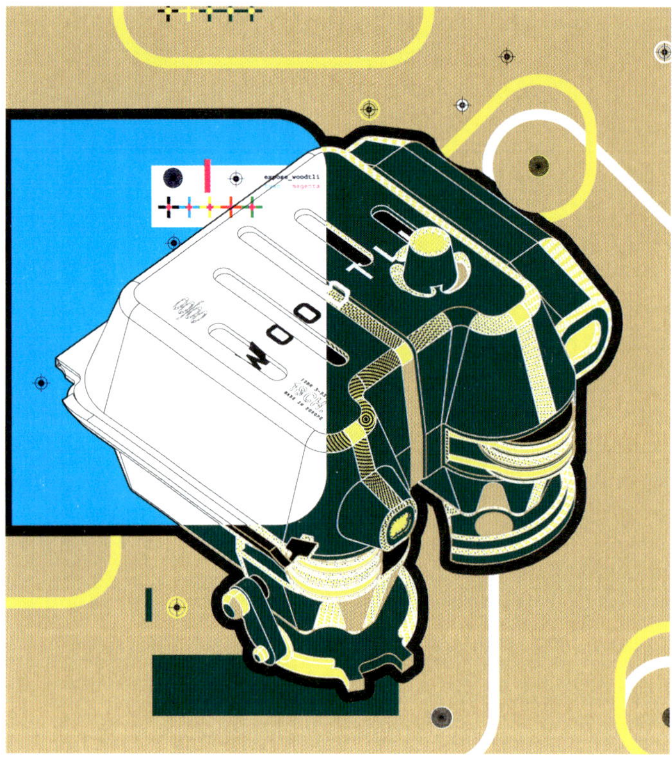

rect in the usual sense of the word. The Woodtli book does not really document existing works, but is rather a document of itself and of a method which reflects its own conditions. ——— Some of Woodtli's earliest pieces (1996 to 1998) are invitations (printed in A5 format on cardboard) to the "Kiosk" in Berne's Lorraine district, where small exhibitions with trend-setting artists took place. He created a particular symbolic form for each of these invitations which reflected the world of the particular artist. In this way, the no longer extant "Kiosk" could develop a visual identity within a short period of time. ——— His **"Sportdesign" poster**, created for the Museum of Design **Zurich** in **2004**, contains a variety of elements which characterize the world of competitive sports as a whole: the track, the seating on the grandstands, the separation of the active and passive participants, the large "S" as a hidden dollar sign, perhaps even the one-dimensional nature of the competitive principle and the possibility to take the whole subject of sports as part of reality, or with the same justification, as bizarre nonsense.

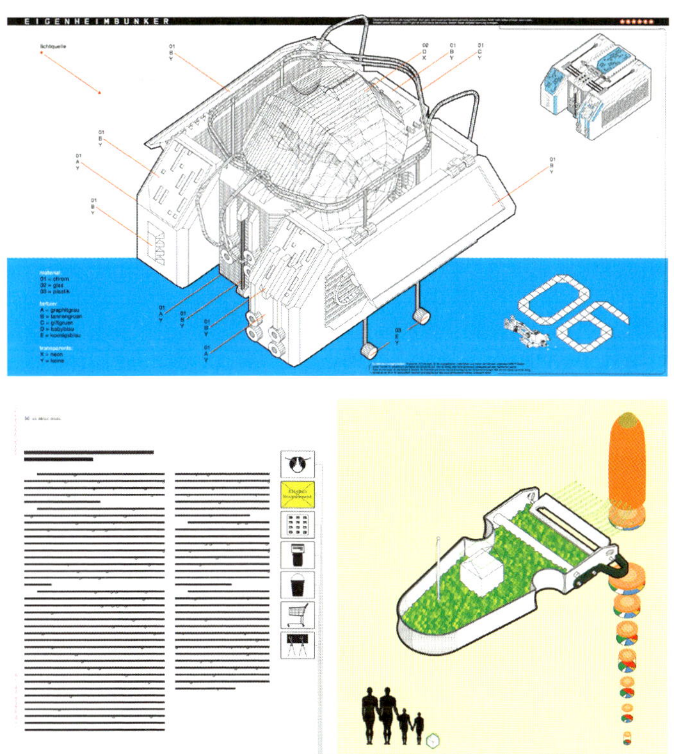

Martin Woodtli, exhibition poster “Sportdesign”, Zurich, 2004

The Bündner Herrschaft is known for its good vineyards. A new building for the elegant **Gantenbein Winery** was built in the town of **Fläsch** in **2006** that likely will become a model of architectural collaboration. Architects **Valentin Bearth** and **Andrea Deplazes**, designers of a number of significant buildings that have been realized in the canton of Graubünden and elsewhere, enlisted the cooperation of their colleagues **Fabio Gramazio** and **Matthias Kohler**, who hold chairs in Architecture and Digital Fabrication at the Swiss Federal Institute of Technology (ETH) Zurich. ——— The design by Bearth and Deplazes called for a concrete skeleton construction, slender in the profile of its members, with floor slabs and perimeter columns visible on the exterior of the building, which was to divide the gentle slope of the property. The infill walls of the high upper level of the story frame structure were required to allow air to circulate and the interior climate to be regulated by natural means. It was the question of the interpretation of this

Gantenbein Winery, Fläsch, just before completion in the late summer of 2006

assignment which led to the cooperation between the two architectural firms. They developed a solution that is exciting in terms of design, method and technology. The large infill panels consist of solid bricks set in their courses at varying angles in a prescribed composition such that some of the joints between the bricks remain open to permit air to circulate and light to penetrate the building's interior. Each brick's orientation was vectorized according to a computer-modeled design intended to evoke an apparently three-dimensional image of circular forms (grapes, perhaps?) through the pattern of the bricks and the resulting play of light and shadows. Each individual brick in the field has a function similar to that of the matrix dot in a printed image. ——— The brick infill panels were made in the summer of 2006 at ETH Zurich. The bricks are bonded by means of a special adhesive which makes the panel construction resistant even to tension forces. The entire production process for the 300-m facade was developed at ETH

Axonometric drawing of a wall infill panel

Robot in action

Placement of the prefabricated wall infill panel

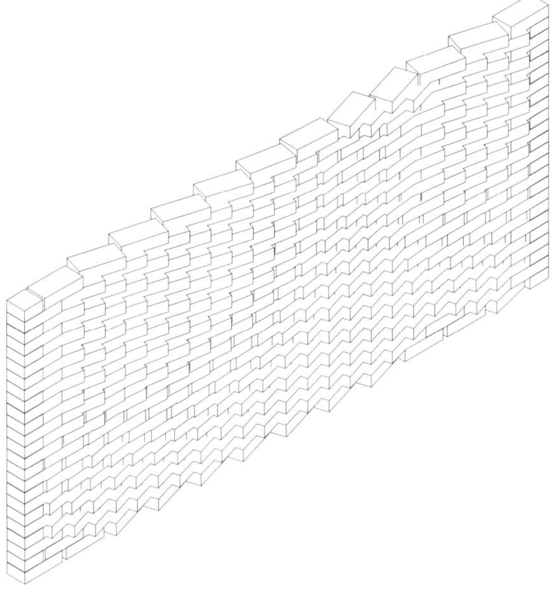

on the basis of initial trials in automated wall-building conducted that spring. An industrial robot constructed the panels in a fascinating spectacle: It grasped and turned each brick according to the programmed command sequence, applied precisely calculated lines of adhesive at specified angles to its underside and placed it in its exact position – brick by brick, course by course, panel by panel. Then the finished infill panels were transported from Zurich to Fläsch, where they were installed ——— This computer-controlled process was employed with the goal of an extremely individualized result. By digitizing the design drawings of the facades and converting them to the scale of the bricks, they were produced materially and with relentless precision. The process makes a significant contribution to the "iconic turn", here not in the sense of the image replacing the word, but of the expected flat wall surface or painted decoration giving way to a three-dimensional material image – to "informed matter", in the words of the architects.

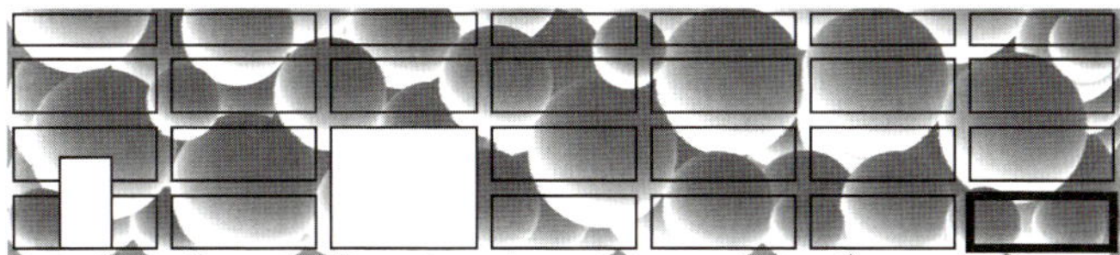

Interior of the storage room

Design drawing, exterior wall

"Save *these* logos before they disappear because of *these* logos." This call to action appears on **www.loslogos.org**. , a virtual project of the **Büro Destruct** studio in Berne **since 2002**, is a response to the ubiquitous inundation with logos and graphics of global companies which is causing the disappearance of an enormous variety of local and regional graphic idioms at a breakneck pace, unnoticed and rarely lamented. **Lorenz Gianfreda**, a founder of Büro Destruct, is trying to counteract this loss. He is defending an unnoticed reality, a cause which has no lobby. Los Logos is a non-commercial project and calls attention to an art which has no real home in the gray area between private influence and public interest. The project aims to protect or at least document cities' own graphic identities. Previously, the German expression *Weich-*

bild (referring to the area within a city's precincts) was used to describe the definite and indefinite elements which made a particular city unique. Büro Destruct aims to highlight that this semi-abstract medium of logos and typefaces must also be considered part of a city's *Weichbild*. ______ Visitors who log on to the website are invited to submit to Büro Destruct photographs of logos and scripts with details on where and when they were taken. Büro Destruct's virtual city matrix is based on square blocks and shows a city at night in perspective. Each block is controlled by a loop function, which randomly selects the photographs to be viewed, the most recent ones being featured and then moving to the background as more are submitted. This technique vividly demonstrates the displacement process.

In central Switzerland, in the town of **Emmenbrücke** near Lucerne, stands a double **studio/residence** which gives an air of Californian effortlessness without resorting to extravagant means. Architect **Jürg Graser** created a captivatingly original design. The long rectangular ground plan of the building corresponds in shape and location to the farmhouse which formerly stood on this site. The **Huber/Blum house** (Judith Huber, Adi Blum), built in **2003**, however, is not a nostalgic building intended to replace the previous one and having to meet the requirements of historical preservation, but instead a refreshing new interpretation of the traditional farmhouse, with living and working quarters under one roof. Two units, each nearly square and housing a living space and a studio, are joined by an open connecting interior court which also serves as a meeting space. In the vertical dimension the units are offset by one story; the south side with its open carport is three-storied, the north side has only two stories. The longer side walls of the 20 m by 282

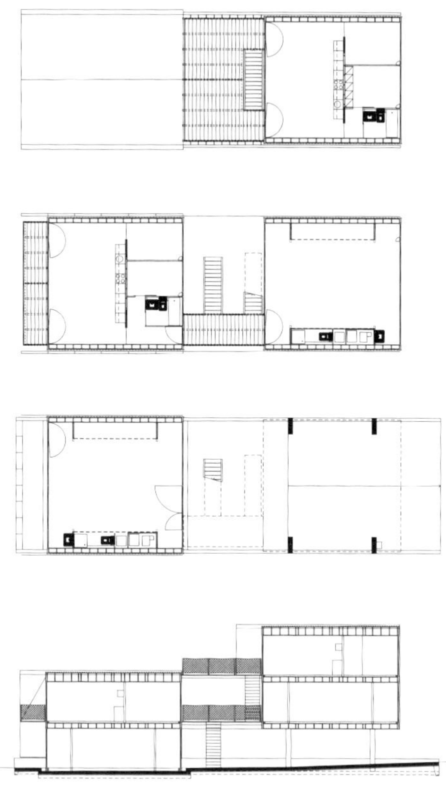

280/281 Huber office building viewed from the street

Floor plans ground floor, second, and third floors

Longitudinal section

View of the inner courtyard

Photograph at night

View from the garden

Photograph of the end wall at night

284 Detail of the enclosure of space

8 m building are closed, the four separate living or working areas are aligned with the longitudinal axis and are lit only from the generously fenestrated end and interior court walls. The spatial, functional, and visual connections across the central hall are unusually varied and rich. ——— This quality is accentuated by the slightly transparent, tent-like fabric covering which stretches over the entire length of the building, connecting the side walls as if it were a roof. Underneath, the supporting structure of glue-laminated wood beams and panels form the horizontal and vertical surfaces (floors and walls), thus defining the interior space. "Walls", however, is not quite the proper term, as the sides of the building are tilted inwards slightly. The stairs and landings in the central hall are constructed of metal grating which lets the light through. Combined with the intentional "encroachments" of the fabric skin spanned across the spaciousness of the inner courtyard, this creates a fascinating flux of shielding and permeability.

BIOGRAPHIES

Aurel Aebi, architect and designer, was born in Biel in 1962. Studied architecture and design at School Athanaeum Architecture & Design, Lausanne, graduated in 1990. Founded the Atelier Oï in La Neuveville with Armand Louis and Patrick Reymond in 1991. Many works in the fields of architecture, design, and scenography; clients include: Ikea, Wogg, Roethlisberger, Swatch, B & B Italia, Expo 02/La Neuveville, Rosenthal, etc. Teaches at the University of art and design Lausanne. Page 260

Bernhard Aebi, architect, was born in Langnau in 1963. Studied architecture at the School of Architecture, Civil and Wood Engineering HSB Burgdorf, graduated in 1988. Staff member of the Atelier 5 in Bern, 1989–1996. Founded the architecture firm Aebi & Vincent in Bern with Pascal Vincent in 1996. Public and private projects: housing developments, single family houses, various renovation projects, Valient Bank in Bern, renovation of the Federal Building (Bundeshaus), etc. Page 194.

Atelier Oï: see Aurel Aebi, Armand Louis and Patrick Reymond.

Les Ateliers du Nord: see Antoine Cahen and Claude Frossard.

Peter Bäder, graphic designer, was born in Schleitheim (Schaffhausen) in 1957. Studied structural engineering in Schaffhausen, 1975–1978; preparatory course and study of graphic design at the Kunstgewerbeschule (School of Arts and Crafts) Zurich, 1978–1982. Since 1983, he has designed album covers and concert, film, and theater posters and was also responsible for concept and design of magazines and newspapers (Strapazin, WOZ, Fabrikzeitung). DTP since 1990. Concept and design of websites and CD-ROMs (RECREC). Taught at the School of Art and Design, Zurich, 1990–2001 and at the State Academy of Design Karlsruhe, 1998–2001. Lecturer at the Lucerne School of Art and Design (graphic design, illustration) since 2001. Page 200.

Rico Baltensweiler, born in Arbon in 1920, died in Lucerne in 1987. Apprenticeship in precision engineering; technician training, 1940–1943; studied electro acous tics, 1943–1945; creative activities with post war youth, 1946–1948; employee of an electronics company and technical engineer for the Swiss Bundesbahn (Federal Railway), 1949–1964. Founded Firma Baltensweiler AG, Ebikon, an atelier for the development and production of lamps in 1951. Page 56.

Rosmarie Baltensweiler, born in Üsslingen (Thurgovia) in 1927. Studied interior design at the Kunstgewerbeschule (School of Arts and Crafts) Zurich, 1942–46. Worked for Max Bill from 1946–1949. Founded the jointly owned Firma Baltensweiler AG, Ebikon, an atelier for the development and production of lamps in 1951. Page 56.

Valentin Bearth, architect, was born in Tiefencastel in 1957. Study of architecture at the Swiss Federal Institute of Technology Zurich, graduated in 1983. Worked with Peter Zumthor 1984–1988. Founded partnership firm with Andrea Deplazes in 1988. Professor at the Accademia di architettura, Mendrisio since 2000. Guest professorship at the Universita di Sassari (I) since 2003. Page 275.

Susi Berger, graphic designer and furniture designer, was born in Bern in 1938. A qualified graphic designer, she has collaborated with Ueli Berger since 1969. Founded the production company Susi + Ueli Berger for the production of graphic design, art, design, furniture design, and architecture. Lives and works in Ersingen. Page 120.

Ueli Berger, painter, sculptor, draftsman, and installation artist, was born in Bern in 1937. Trained as a painter; worked for various architects, including Hans Brechbühler. Industrial designer and member of the avant-garde group "Bern 66", 1965–1969. Since 1969, collaboration with his wife, Susi Berger. Regular contributor to the Swiss Plastikaustellung (Sculpture Exhibit) in Biel, 1966–1986. Since 1972, artistic overall design of various buildings and public spaces – interventions and performances. Lecturer (plastic arts) at the Swiss Federal Institute of Technology Zurich, 1981–1986. Co-founded the class for further education in the fine arts at School of Visual Arts Bern and Biel in 1987, which he taught until 1994. Lives and works in Ersingen. Page 120.

Max Bill, architect, designer, fine artist, and author, was born in Winthherthur in 1908, died in Berlin in 1994. Studied silversmithing at the Kunstgewerbeschule (School of Arts and Crafts Zurich); studied at Bauhaus Dessau in 1929. After his return to Switzerland he took part in the avant-garde movement: as architect, pioneer of modern commercial art, and member of the group abstraction-création. He worked as an industrial designer, was a leading constructivist artist and initiated both the traveling exhibition Die gute Form (The good form) in 1949 and, later, the Schweizer Werkbund award of the same name (1951–1968). Co-founder, designing architect and first president of the Ulm School of Design (1951–1956). Architect in charge of the "bilden und gestalten" (building and design) section of the Swiss pavilion at the 1964 Expo in Lausanne. Pages 38,40,61.

Dimitri Bruni, graphic designer, was born in Biel in 1970. Studied at the School of Visual Arts Bern and Biel, graduated in 1996. Founded the agency Norm with Manuel Krebs in 1999. Norm designs and publishes written works and books. The book publications include research projects initiated by the agency in the area of graphic design and typographic design, the two most important examples are the books *Norm: Einführung* ("Norm: Introduction") and *Norm: Die Dinge* ("Norm: The Things"). Examples of typographic design include "Simple", designed for Cologne-Bonn airport, and "CT-Omega", for the Swiss watchmaker of the same name. Page 250.

Susanne Brütsch, designer in training, was born in Schaffhausen in 1981. After completing secondary school in Schaffhausen, she lives in Biel since 2001 (preparatory course in design). Gained admittance to the industrial design department of the University of Applied Sciences Northwest Switzerland in 2004, where she still studies. Page 254.

Sibylle Bucher, architect, was born in Zurich in 1965. She studied architecture at the Swiss Federal Institute of technology Zurich, internships with Swiss and foreign architects (Lugano, Barcelona), graduated in 1990. Worked with Bétrix and Consolascio Architects (Erlenbach) and with Burkhalter and Sumi (Zurich). Architecture firm with Christoph Elsener in Zurich and Rorschach, 1994–1996; founded B.E.R.G. Architects with Christoph Elsener and Michel Rappaport in Zurich in 1995. Page 252.

Hani Buri, architect, was born in Bern in 1963. Studied architecture at the École Polytechnique Fédérale de Lausanne, graduated in 1991. Intern with Dunning & Versteegh in Geneva and with Daniel Liebeskind while still at university. Freelance architect after graduated, founded the firm bmv in Geneva with Olivier Morand and Nicolas Vaucher in1993. Assistant Professor at the École Polytechnique Fédérale de Lausanne. Page 202.

Franziska Burkhardt, graphic designer, was born in Zurich in 1972. Studied at the School of Art and Design Zurich; freelances in Zurich under the name Buka Grafik since 1996. Collaborations with Nadine Spengler and Anna Albisetti, among others. Page 268.

Büro Destruct: see Lorenz Gianfreda.

BMV Architectes: see Nicolas Vaucher, Hani Buri, and Olivier Morand.

Gion A. Caminada, architect, was born in Vrin in 1957. Trained as a carpenter, he visited the School of Arts and Crafts and received his qualification as an architect from the Swiss Federal Institute of Technology Zurich. Founded his own architecture firm in Vrin. Assistant professor at the Swiss Federal Institute of Technology Zurich since 1998. Projects, esp. in Vrin and vicinity: stiva da morts (mortuary), boarding school for young women in Disentis, among others. Page 182.

Antoine Cahen, industrial designer, was born in Lausanne in 1950. Studied industrial design at the University of art and design Lausanne, graduated in 1975. Founded his own studio in 1975 while also teaching at the University of art and design Lausanne. Founded the Ateliers du Nord (Studio for Industrial Design) in Lausanne with Claude Frossard and Werner Jeker in 1983. His brother Philippe Cahen joined the team in 1996. Several products including concept and design for Nespresso, flashlights, Logitech etc. Page 248.

Andreas Christen, designer und fine artist, was born In Bubendorf in 1936, died in Zurich in 2006. Studied privately with Hans Fischli at the Kunstgewerbeschule (School of Arts and Crafts) Zurich. Founded his own design studio in 1960; work for 64 Expo in Lausanne, collaboration with the firm of R. Lehni (until 2006); collaboration with Knoll International since 1967; taught at the Hamburg College of Fine Arts 1971–1972. President of the Swiss Federal Commission of Fine Arts and Design, 1989–1996. Page 114.

Willi E. Christen born in Murgenthal (Aargau) in 1935. Architect and architectural draftsman, staff member of the architecture firm Zweifel & Strickler (Zurich), 1959–1964: Expo 64 in Lausanne. Founded his own firm in Zurich in 1965. Several public and private projects/agricultural buildings in Switzerland including: the Juchhof, city estate of Zurich, the St. Luzisteig military quarters, the Agricultural College Giswil 1973; the University Zurich-Irchel (main library, new and building alterations to the zoology and veterinary departments, and several private residences). Managing Director of the trade journal "Werk, Bauen + Wohnen" 1980–1999, taught at the Swiss Federal Institute of Technology Zurich 1988-2000. Editor of the Schweizer Architekturführer (Swiss Achitecture Guide) from 1920 to 1995. Lives in Zurich. Page 122.

Roman Clemens, artist and set designer, born in Dessau in 1910, died in Zurich in 1991. Studied at Bauhaus Dessau 1927–1931, with Kandinsky, Klee, and Mies van der Rohe among others; works for the Bauhaus stage. Settled in Zurich in 1932, set designer at the Zurich Stadttheater until 1943, stage design for over 125 productions. Interior design of the cinema Studio 4 in Zurich with architect W. Frey in 1949. Also worked as a constructivist painter. Recipient of the medal of honor of the Canton of Zurich in 1990. Page 22.

Patricia Collenberg, textile designer, was born in Chur in 1968 Chur. Studied textile design at the Zurich School of design, 1993–1997, graduated in 1997. Employment in the industry, 1997–1999, freelance work for several companies since then. Occasionally teaches at the School of Art and Design Zurich. Collaboration with Zuzana Ponicanova under the name Collenberg/Ponicanova since 2000. Lives and works in Zurich. Page 226.

Jürg Conzett, civil engineer, was born in Aarau in 1956. Studied at the Swiss Federal Institute of Technology Zurich and the École Polytechnique Fédérale de Lausanne, graduated in 1980. Collaboration with Oeter Zumthor, 1981–1987. Founded his own firm in Haldenstein (Grisons) in 1988. Collaboration with A. Branger (Melcherts + Branger AG) in Chur, first under the name Branger & Conzett AG (1992–1996), then under the name Conzett, Bronzini, Gartmann AG (since then). Lecturer at the University of Applied Sciences HTW Chur since 1985. Multifarious activities including: project planning and site management for bridges and buildings, historic preservation research, protection and restoration, etc. Diverse publications. Page 228.

Heinrich Danzeisen, architect, was born in Degersheim (St. Gallen) in 1919. Studied at the Technikum Winterthur (now Zurich University of Applied Sciences), graduated in 1944. Founded the architecture firm Danzeisen und Voser (Hans Voser, architect, born in Effretikon (Zurich) in 1919, died in St. Gallen in1992) in 1950. The firm worked primarily in the St. Gallen area and performed all important architectural tasks. Danzeisen focused on the development of design of industrial buildings. Many pioneering projects, among others: factory hall for the Elastic AG in Gossau, carpet housing Biserhof St. Gallen. Page 32.

Bernard Delefortrie, architect, was born in Uccle (Belgium) in 1959. Studied architecture at the School of Architecture St. Luc in Brussels, graduated in 1982. Employed in western Switzerland 1983-1986, at the same time working for Groep Planning, Brussels (project management in Morocco); First freelancer and then employee at Claude Rollier SA (Neuchâtel) and at Ch. Feigel (Auvergnier), 1986-1994. Self-employed since 1994; since 1995, partnership with Laurent Geninasca. Page 246.

Andrea Deplazes, architect, was born in Chur in 1960. Study of architecture at the Swiss Federal Institute of Technology Zurich, graduated in 1988. Founded partnership firm with Valentin Bearth in 1988. From 1997 onwards professor of architecture and construction at the Swiss Federal Institute of Technology Zurich, since 2005 dean of the architecture department at the Swiss Federal Institute of Technology Zurich. Page 275.

Roger Diener, architect, was born in Basel in 1950. Studied architecture at the Swiss Federal Institute of Technology Zurich, graduated 1975. Founded the firm Diener & Diener together with his father Marcus Diener (b. 1918 in Basel, managed his own firm since 1942) in 1976. Developed an experimental designing procedure akin to inquisitive research. The firm takes on a wide variety of contracts: multiple family dwellings/apartment buildings, office buildings, schools, museums, hotels and businesses/ industrial areas. Participant in numerous national and international competitions. Page 240.

Christoph Dietlicher, designer, was born in Zurich in 1958. After training as an elementary school teacher and some years of teaching, he studied jewelry and tools at the Zurich School of Design, graduated 1988. After graduation, founded the Neue Werkstatt (New Studio) with Thomas Drack and Andreas Giupponi, remaining a partner until this day. Lecturer at the School of Art and Design Zurich (industrial design) since 1990. Lives in Zurich. Page 152.

Thomas Drack, designer, was born in Zurich in 1962. After training as a mechanic, he taught himself metal furniture design. Founded the Neue Werkstatt (New Studio) with Christoph Dietlicher and Andreas Giupponi in 1998, left in 2005. Lives in Winterthur. Page 152.

Harald Echsle, architect, was born in Davos in 1968. Study of architecture at the Swiss Federal Institute of Technology Zurich, graduated in 1997. Founded partnership firm with Annette Spillmann in Zurich in 2002. Page 266.

Christoph Elsener, architect, was born in Rorschach (St. Gallen) in 1964. Studied architecture at the Swiss Federal Institute of Technology Zurich, several internships, graduated 1990. Joint firm with Sibylle Bucher in Zurich and Rorschach, 1994–1996; since 1995 joint firm (B.E.R.G. Architects) with Sibylle Bucher and Michel Rappaport in Zurich. Teaches architecture and construction at the Swiss Federal Institute of Technology Zurich. Page 252.

Helmut Federle, artist, was born in Solothurn in 1944. Studied at the Kunstgewerbeschule (School of Arts and Crafts) Basel, 1964–1969, traveled extensively abroad during that period. First individual show at the Riehentor Gallery in Basel in 1971; visited the Cité Internationale des Arts Paris, 1971–1972; moved to New York in 1979; studio in Geneva and teaching position at the École des Beaux-Arts, 1981; lived in Zurich from 1983–1984, then moved to Vienna. Several Exhibitions, collaborations include his work on the "Forum 3" building on the Novartis Campus (with Diener & Diener and Gerold Wiederin). Page 240.

Beat Frank was born in Bern in 1949. After training as a graphic designer he worked in advertising agencies, Young & Rubicam among others, and as an illustrator for scientific publications. Founded his own studio Atelier Vorsprung for the design of three-dimensional objects in 1984. Contributed to the Documenta in 1987. Several exhibitions in Germany, Italy, and Switzerland. Teaches at the School of Visual Arts Biel since 1985. Page 163.

Claude Frossard, architect, born in Ardon (Wallis) in 1947. Training as an architectural draftsman in Lausanne 1964–1968. Employed by diverse architecture firms 1968–1971 . Study of Industrial Design at the University of art and design Lausanne, graduated in 1976. Freelance industrial designer 1976–1979. Worked on assignment as an industrial designer in the Karachi Design Institute Pakistan for the EDA (Swiss Federal Department of Foreign Affairs) /development aid agency. From 1979 to1982 he traveled through Asia while simultaneously working in collaboration with Antoine Cahen. Founded the partnership Ateliers du Nord (a firm for industrial design) in Lausanne with Antoine Cahen and Werner Jeker in 1983. Several productions, including: Nespresso (concept and design), flashlights, Logitech, scenography assignments etc. Page 248.

Franz Füeg, architect, was born in Solothurn in 1921. After training as an architectural draftsman with Robert Winkler (architect 1898–1973) in Zurich, he founded his own firm in Solothurn in 1953. Lives in Zurich since 1975, collaboration with Melchior Wyss since the 1980s, partners in the firm Franz Füeg, Melchior Wyss Architekten since 1991. Editor of "Bauen + Wohnen" 1958–1961; visiting professor at several foreign universities; professor at the École Polytechnique Fédérale de Lausanne 1971–1987. Broad range of projects including: homes, businesses, schools and institutions, churches (Catholic Church St. Pius in Meggen), and Museums (renovation of the Kunst-museum Solothurn); several research contracts. Page 118.

Jürg Fontana, architect and designer, was born in Zurich in 1967. Trained as an architectural draftsman in Zurich, studied architecture at Technikum Winterthur (now Zurich University of Applied Sciences) in 1988, graduated in 1992. Four years employed as an architect with Martin Spühler; since 1995, self-employed as designer and producer, particularly of lights and lamps. Page 238.

Andreas Fuhrimann, architect, was born in Zurich in 1956. Studied physics (4 semesters) and architecture at the Swiss Federal Institute of Technology Zurich, graduated in 1985. One year as drafter and planner in the architecture firm Marbach + Rüegg. Collaboration with Christian Karrer since 1987. Lecturer at the School of Arts and Crafts (interior design) since 1988. Collaboration with Gabrielle Hächler since 1995. Page 232.

Aurelio Galfetti, architect, was born in Biasca (Ticino) in 1936. Studied architecture at the Swiss Federal Institute of Technology Zurich, 1954–1960. Founded his own architecture firm in Lugano in 1960. Collaborations with Flora Ruchat, Livio Vacchini, Luigi Snozzi, Rino Tami, and Mario Botta among others. Several public and private projects, esp. in Bellinzona: restoration of the castle 1981–1991), residential houses, sports complex; design of the Bernina Pass dam (1991); planning for the new railway crossing of the Alps, NEAT. Page 133.

Laurent Geninasca, architect, was born in Neuchâtel in 1958. Studied architecture at the Swiss Federal Institute of Technology Zurich, graduated in 1984. Internships with Giovanni Michelucci (Florence), Antoine Grumbach (Paris), and Luigi Snozzi (Locarno), 1979–82. Managerial position at the 1991 architecture exhibition on the 700th anniversary of the Swiss Confederation. Self-employed, 1991–1994; 1994, participation in the design entry for the 02 Expo Swiss pavilion. Visiting professor in Mendrisio. Partnership with Bernard Delefortrie since 1995. Page 246.

Karl Gerstner, typographer, graphic designer, and artist, was born in Basel in 1930. Learned graphic design with Fritz Bühler (Basel); studied at the Kunstgewerbeschule (School of Arts and Crafts) Basel (Armin Hofmann, Emil Ruder). Employee of the J. R. Geigy studio (Basel) since 1949. Founded a studio for Werbung, Grafik, Publizität (advertising, graphic design, publicity) with Markus Kutter in Basel 1959 (called GGK since 1961). Retreat from advertising in 1970; work as a fine artist and as the author of books on graphic design and culinary art. Lives in Basel and in Alsace. Page 88.

Lorenz Gianfreda, graphic designer, was born in Bern in 1971. Preparatory course at the School of Visual Arts Bern and Biel; training as a graphic designer in Kurt Wirth's studio. Founding member of The Büro Agency, Bern. Page 278.

Ivano Gianola, architect, was born in Biasca in 1944. Attended grammar school in Ticino, trained as an architectual draftsman in Zurich 1962-1965. Followed by self-taught study of architecture. Founded his own architecture firm in Mendrisio in 1969. Page 150.

Annette Gigon, architect, was born in Herisau (Appenzell Ausserhoden) in 1959. Studied architecture at the Swiss Federal Institute of Technology, Zurich, graduated in 1984. Staff member of the architecture firm Marbach und Rüegg in Zurich and Herzog & de Meuron in Basel, 1984–1987. Founded her own firm in Zurich, 1987–1989; partnership with Mike Guyer. First joint project: Kirchner Museum in Davos (1989–1992). Several public and private projects since then, including: Restaurant Vinikus Davos, extension of the Kunstmuseum Winterthur, sports complex Davos, single family homes, housing developments. Page 218.

Ernst Gisel, architect, was born in Adliswil (Zurich) in 1922. After training as an architectural draftsman, employed by Alfred Roth, studied at the Kunstgewerbeschule (School of Arts and Crafts) Zurich, 1940–42. Founded his own firm in 1945, partnership with Ernst Schaer until 1947. Won many competitions early on, beginning with the Park-theater Grenchen (1949). Many public and private projects in Switzerland and in Germany, including: residential homes (from single family houses to housing developments), administrative buildings and ecclesiastical architecture, schools, swimming pools, theaters. Lives in Zurich. Pages 96, 98, 100.

Andreas Giupponi, designer, was born in Davos in 1958. After training as a precision engineer, he studied jewelry-making and tools at the School of Art and Design Zurich, graduated in 1984. Designed jewelry until 1988, when he founded the Neue Werkstatt (New Studio) with Christoph Dietlicher and Thomas Drack, where he remains a partner. Lives in Zurich. Page 152.

Fabio Gramazio, architect, was born in Langenthal (Solothurn) in 1970. Studied architecture at the Swiss Federal Institute of Technology Zurich, graduated in 1996. Co-founded the art project etoy, 1994–2000. Teaching and research (Architecture and CAAD) at the Swiss Federal Institute of Technology Zurich, 1996–2000. Partner in the firm Gramazio & Kohler since 2000. Assistant Professor at the Swiss Federal Institute of Technology Zurich since 2005. Page 275.

Jürg Graser, architect, born in Bern in 1965. Studied architecture at the Swiss Federal Institute of Technology Zurich, graduated in 1991. Staff member of the architecture firm M. Caroline und Olivier Tissier/Pierre Caillot, Paris, 1992–1995; founded architecture firm with Christian Wagner in Sargans and Zurich, 1995–2000; has his own firm in Zurich since 2000; taught at the University of Applied Sciences HTW Chur, 1995–2000, research on the "Jurasüdfuss-Architekten" (Jura Mountain Architects) at the Swiss Federal Institute of Technology Zurich since 2000. Page 280.

Christian Guggenbühl, hand bookbinder and designer, was born in Zurich in 1961. Trained as a hand bookbinder, including studying at the Kunsgewerbeschule (School of Arts and Crafts) Zurich, 1977–1980. Employed by several bookbinders, including Burkhardt, Zurich; master bookbinder since 1987, took over a business in 1993 (Buch-Atelier Rohrer) and founded Format Guggenbühl. Expanded the business to include product design, exhibition systems, and exhibition design. Founded Simply FG in 2005 to market his own products. Page 239.

Willy Guhl, interior designer and product designer, was born in Stein am Rhein in 1915, died in Hemishofen (Schaffhausen) in 2004. After training as a cabinetmaker (1930–1933), he studied interior design at the Kunstgewerbeschule (School of Arts and Crafts) Zurich (with Wilhelm Kienzle). Founded his own studio in Zurich in 1933; from 1941–1980 he taught interior design at the Kunstgewerbeschule (School of Arts and Crafts) Zurich, department head from 1951 onwards. Groundbreaking work in Swiss interior design, furniture and product design, productive industrial collaboration (with Eternit AG, among others); design and production of many pieces of furniture (including living room furniture, storage units, the Guhl-stool, the Scobalit shell chair, Eeternit lounge chairs, etc.), multifunctional machines (tractor, Mehrzweckmaschine Aebi KM 54). Page 62.

Mike Guyer, architect, was born in Columbus, Ohio (USA) in 1958. Studied architecture at the Swiss Federal Institute of Technology Zurich, graduated in 1984. Staff member of the architecture firm OMA in Rotterdam, 1984–1987; in Zurich since then, had his own firm until 1989, Assistant to the Hans Kohlhoff chair at the Swiss Federal Institute of Technology Zurich, 1987–1988, founded a partnership with Annette Gigon in 1989, first joint Kirchner Museum in Davos (1989–1992). Several public and private projects since then, including: Restaurant Vinikus Davos, extension of the Kunstmuseum Winterthur, sports complex Davos, single family homes, housing developments. Page 218.

Gabrielle Hächler, architect, was born in Lenzburg (Aargau) in 1958. Studied Art History at the University of Zurich and architecture at the Swiss Federal Institute of Technology Zurich, internship with Prof. Dolf Schnebli, awarded a scholarship for a semester abroad at the School of Architecture in Ahmedabad, India, graduated in 1988, student of Prof. Mario Campi, Assistant teacher for 4 years (construction, Lecturer: Ivo Trümpy) at the Swiss Federal Institute of Technology Zurich. Founded his own firm in 1988; temporary collaborations with other architects and artists, collaboration with Andreas Fuhrimann since 1995. Page 232.

Christoph Haerle, sculptor and architect, was born in Zurich in 1958. Trained in stone sculpture-making and then studied architecture at the Swiss Federal Institute of Technology Zurich, graduated in 1987. Freelancing artist since 1983, studio work with Daniel Kündig, Daniel Blick and Sabina Hubacher, 1984–1991. Joint studio with Sabina Hubacher since 1996. Several teaching positions, jobs requiring expertise and exhibitions. Page 240.

Fritz Haller, architect, was born in Solothurn in 1924. Training as an architectural draftsman, employee in several architecture firms in Switzerland. Architect in Rotterdam with van Tijen and Maaskant, 1948. Self-employed architect in Solothurn since 1949, until 1962 together with his father Bruno Haller. Several teaching positions in Switzerland and abroad (Professor in Karlsruhe). Many years collaboration with the company USM (Münsingen). Lives in Bern. Page 102.

Thomas Hasler, architect, was born in 1957. First trained as a cabinetmaker, then studied architecture at the Technikum Winterthur (now Zurich University of Applied Sciences), followed by a study of architecture at the Swiss Federal Institute of Technology Zurich, graduated in 1989. Wrote his dissertation on the German architect Rudolf Schwarz in 1997. Founded a partnership architecture firm with Astrid Staufer in Zurich and in Frauenfeld in 1994. Page 198.

Walter Henne, architect in Schaffhausen, died in 1990 (no further data available). Page 58.

Hans Hilfiker, architect/engineer, was born in Zurich in 1901, died in Gordevio (Ticino) in 1993. Trained as a precision mechanic in Zurich, then studied electrical engineering at the Swiss Federal Institute of Technology Zurich, graduated in 1925. Project planning and management for laying telecommunication cables in Argentina for Siemens-Werke Berlin. Returned to Switzerland in 1931, engineer at the Swiss Federal Railway (SBB) (Bauabteilung III construction section III), 1932; associate director of the Bauabteilung and head of operations of the Swiss Federal Railway, 1944. Several projects and buildings including the design of a new loading crane, 1950; design of a convertible street/track vehicle for railway track maintenance, 1951; design of the SBB railway clock (still the same today), 1955; design and construction of the station platform roves at Winterthur-Grüze, 1952–1955, built-in kitchen SINK-Norm. Page 42.

Armin Hofmann, graphic designer, was born in Winterthur in 1920. Studied at the Kunstgewerbeschule (School for Arts and Crafts) Zurich; trained as a lithographer in Winterthur, 1939–1943; employed as a lithographer 1943–1948, founded his own studio in Basel and Lucerne in 1948. Graphic design teacher at the Basel School of Design, 1946–1986; director of the further education course in visual communication at the School of Design since 1986; frequent guest lecturer at American universities and at the National Institute of Design in Ahmedabad, India since 1955. Works of public art and posters for museums and theaters in Basel, several publications. Lives in Lucerne. Page 77.

Ho-La: see Jacqueline Lalive d'Epinay and Sabine Leuthold.

Heinz Hossdorf, civil engineer, was born in Wiesbaden in 1925, died in Madrid in 2006. Studied civil engineering at the Swiss Federal Institute of Technology Zurich, left before graduating. Founded his own studio with a pilot project laboratory in 1953; collaborations with architects Danzeisen & Voser and Otto & Walter Senn (University Library, Basel), Schwarz & Gutmann (Stadttheater Basel) among others. Pioneer user of the computer for design and realization of supporting structures. Lived in Basel and, since 1984, Madrid. Pages 32, 100.

Sabina Hubacher, architect, was born in Zurich in 1957. Preparatory course at the Kunstgewerbeschule Zürich (School of Arts and Crafts Zurich), studied architecture at the Swiss Federal Institute of Technology Zurich, graduated in 1983. Worked with Daniel Kündig, Daniel Bickel, and Christoph Haerle, 1984–1991. and with Brigitte Widmer 1992–1996. Partnership with Christoph Haerle since 1996. Several teaching jobs and jobs requiring expertise. Page 204.

Markus Huber Recabarren, architect and designer, was born in Zurich in 1969. After training as an architectural designer, he studied interior design and product design at the Zurich School of Design. Exchange student at the School of Visual Arts, NYC, graduated in 1997. Founded Zenaro GmbH for architecture and design with Michel Nigg in 1997 (until 2006). Several renovations and interior design commissions: restaurants, salesrooms, day care centers, exhibitions, corporate design, and furniture design. Designed the shelving unit "Plattenspiel" in 1999. Architect and product designer for Pipilotti Rist, Zurich, since 2004. Page 216.

Heinz Isler, civil engineer, was born in Zollikon in 1926. Studied at the Swiss Federal Institute of Technology Zurich, graduated in 1950. Assistant (to Prof. Lardy) for static calculation and solid building at the Swiss Federal Institute of Technology Zurich 1950–1953. Then studied at the Kunstgewerbeschule (School of Arts and Crafts) Zurich (for 9 months) and freelanced in different engineering firms. Developed new methods of building thin non-geometric shells, 1954–1955. Participated in the founding conference of the International Association for Shell Structures in Madrid in 1959. Founded his own firm with laboratory and testing ground in Burgdorf in 1956. Design and project planning of many shell structures for industrial sites and sports centers, garden centers, and shopping centers, residential homes, churches, theaters, museums, etc. Works in Burgdorf, lives in Zuzwi (Bern). Pages 104, 110.

Renate Jaberg, graphic designer, was born in Bern in 1972. Preparatory and specialized graphic design courses at the School of Visual Arts Bern and Biel; Internships in Berlin (Meta Design) and Amsterdam (Total Design). After employment at Éclat Design agency Erlenbach, work as a freelance designer. Several works for design agencies. Freelances since 1998 under the name «grafiksalon». Page 224.

Samuel Jäggi, Teacher and freelance designer, was born in Bern in 1977. Training as a group teacher with a focus on the artistic design, 1995–2000. During his training, part-time employment as a landscape gardener. Teaches design and drawing in Konolfingen since 2001. Founded the company Gestaltsam in 2003 and launched a velcro textile project. Several technical works and the production of different pieces of furniture. Page 214.

Werner Jeker, graphic designer, was born in Mümliswil (Solothurn) in 1944. Training with Hugo Wetli in Bern. Founded his own studio in 1972; collaborations with various cultural institutions; art director of L'Illustré (1980–82). Joint studio Les Ateliers du Nord' with Antoine Cahen and Claude Frossard in Lausanne since 1984. Corporate Design for Weimar Cultural City of Europe 1999, (Art Director of Communication); architecture/scenography for the Pavilion Signalschmerz, Expo 02. Several teaching appointments in Switzerland and abroad (Lausanne/Paris/Karlsruhe), Co-director of Visual Communication Studies at the Bern University of the Arts since 2003. Lives and works in Lausanne and Bern. Page 160.

Felix Jerusalem, architect, was born in Freiburg, Germany in 1963. Moved to Zurich in 1968, lived in the USA (Rochester/Minnesota) 1973-1974. Studied architecture at the Swiss Federal Institute of Technology Zurich, graduated in 1991. Several internships in Switzerland and abroad: 1991 with Bernard Tschumi in New York, 1991–1998 with Ernst Gisel in Zurich. Founded his own firm in Zurich in 1995, several projects. Page 262.

Theres Jörger, visual designer, was born in Laax in 1973. Studied visual design at the Lucerne School of Art and Design, graduated in 2001. Runs her own studio in Zurich, area of particular interest: typography and space. Lives in Zurich. Page 258.

Christian Kerez, architect, was born in Maracaibo, Venezuela in 1962. Studied architecture at the Swiss Federal Institute of Technology Zurich, graduated in 1988. Employed as an architect by Rudolf Fontana, 1991–1993. After publishing many works on architectural photography, he founded his own architecture firm in Zurich in 1993. Visiting professor and Assistant Professor at the Swiss Federal Institute of Technology Zurich since 2001. Page 270.

Wilhelm Kienzle, interior designer and product designer, was born in Basel in 1886, died in Zurich in 1958. Trained as carpenter and furniture designer and took courses at the Basel School of Design, 1901–1905; further education as an ornamental metalworker; self-employed in Munich 1909–1912: designer of trade and industry furniture. interior design, commercial art, sometimes freelancing for others, sometimes for himself. Afterwards, employed in various architecture firms in Germany. He returned to Switzerland in 1916; becoming director of specialized courses for interior design at the Kunstgewerbeschule (School of Arts and Crafts) Zurich (until 1951). Several pieces of furniture and basic commodities, including the collapsible bookshelf (1931), the cactus watering can (1935), One room apartment for bachelors (Exhibition "Das Neue Heim" (The New Home), Kunstgewerbemuseum Zurich (1926)). Page 48.

Matthias Kohler, architect, was born in Uster (Zurich) in 1968. Studied architecture at the Swiss Federal Institute of Technology Zurich, graduated in 1996. Taught and researched (chair for architecture and design) at the Swiss Federal Institute of Technology Zurich, 1997–1999; taught and researched (mechanical processes in architectural design) at the Swiss Federal Institute of Technology Zurich, 1999–2000. Co-proprietor of the firm since 2000. Assistant Professor at the Swiss Federal Institute of Technology Zurich since 2005. Page 275.

Hanna Koller, graphic designer, was born in Meierskappel (Lucerne) in 1966. Preparatory course and graphic design course at Lucerne School of Design 1982–1987. Graphic designer with Seiniger Advertising, Movie Poster Company Hollywood in Los Angeles, CA (USA) in 1988. Graphic designer with ASGS advertising agency in Zurich in 1989. Founded her own graphic art studio in Zurich in 1990. Lives and works in Zurich. Page 236.

Manuel Krebs, graphic designer, was born in Bern in 1980. Studied at the School of Visual Arts Bern and Biel, graduated in 1996. Founded the agency Norm with Dimitri Bruni in 1999. Norm designs and publishes written works and books. The book publications include research projects initiated by the agency in the area of graphic design and typographic design, the two most important examples are the books *Norm: Einführung* ("Norm: Introduction") and *Norm: Die Dinge* ("Norm: The Things"). Examples of typographic design include "Simple", designed for Cologne-Bonn airport, and "CT-Omega", for the Swiss watchmaker of the same name. Page 250.

Jacqueline Lalive d'Epinay, designer and graphic designer, was born in Schlieren (Zurich) in 1974. Studied furniture and product design at Kingston University in London. During her studies, she completed several internships at home and abroad. She graduated in 1997, and returned to Zurich to work for GAAN in the field of 3D design. In 2000, she founded the label Ho-La with Sabine Leuthold (through 2002), and designed and produced several products. Since 1999 further education in the fields of graphic design, communications, and concepts; since then, works with a variety of agencies and institutions. Page 186.

Sabine Leuthold, product designer and interior designer, was born in Bern in 1972. Studied product design at the University of art and design Lausanne, graduated in 1997. Several internships and freelance works with design agencies 1998–2000. In 2000, she founded the label Ho-La with Jacqueline Lalive d'Epinay (through 2002), and designed and produced several products. Founded her own design agency in the fields of product and interior design in 2001. Designs for lamps and other home furnishings with international companies. Workshops with F+F Art School Zurich. Several works for exhibitions and interiors. Page 186

Richard Paul Lohse, artist and graphic designer, was born in Zurich in 1902 and died there in 1988. Training as an advertising artist 1918–1922, self-taught painter. Freelance graphic designer and book designer in the 1930s. Founded the "Allianz, Vereinigung moderner Schweizer Künstler" ("Alliance, Society of Modern Swiss Artists") with Leo Leuppi in 1937. Worked on the exhibition "Twentieth Century German Art" in London in 1938. A political person, he was active in the Resistance. Breakthrough in painting in 1943. Member of the editorial board of "Bauen+Wohnen", 1948–1956, co-editor of the magazine "Neue Grafik", 1958–1965. Participated in numerous exhibitions on systematic-constructive art and constructive commercial art, including Biennale Sao Paolo 1965 and Venice 1972, Documenta 1968, Stedelijk Museum Amsterdam, etc. Page 28.

Ania Losinger, musician and flamenco dancer, was born in Bern in 1970. Studied flamenco in Switzerland and Spain for many years. Studied rhythmics at the Conservatory in Zurich, graduated in 1993. Engaged as a dancer with the dance company Flamencos en route, 1997–1999. Developed the floor xylophone with the inventor and instrument maker Hamper von Niederhäusern, 1998–1999. Intensive performance career with the Xala, both as a soloist and in small formations with Don Li and in the duo "The Five Elements" with Matthias Eser. Guest performances in Europe, the US, and Russia. Page 222.

Armand Louis, designer and naval architect, was born in Biel in 1966. Studied naval architecture in Lausanne, graduated in 1986. Worked as a freelance designer in La Neuveville 1986–1991, several freelance works for design agencies. Founded the agency Atelier Oï in La Neuveville with Patrick Reymond and Aurel Aebi in 1991. Many works in the fields of architecture, design and scenography; clients include: Ikea, Wogg, Roethlisberger, Swatch, B&B Italia, Expo 02/La Neuveville, Rosenthal etc. The agency has a professorship in the Department of Design at the University of art and design, Lausanne. Page 260.

Hans-Rudolf Lutz, typographer and artist, was born in Zurich in 1939 and died there in 1998. Training as a typesetter in the Orell Füssli printing company, Zurich. Worked with typographer Arthur Kümin in 1959. Typesetter in the Anton Schöb printing company, Zurich, in 1960, then travels through Europe and North Africa. Course in typographic design with Emil Ruder and Robert Büchler in Basel in 1963. Head of the group "expression typographique" in the "studio hollenstein" in Paris in 1964. Taught typesetting at the Zurich School of Design in 1966. Founded his own studio in Zurich – Hans-Rudolf Lutz Publishers – Publishing Cooperative – Institute of History and Theory of Architecture – typography teacher and interdisciplinary design in Lucerne since 1968. Guest teaching positions at various schools since 1964. Page 156.

Beat Mathys, architect, was born in Bern in 1962. Studied architecture at the Swiss Federal Institute of Technology Zurich, graduated in 1987. Founded the architecture firm smarch with Ursula Stücheli in 1991. Developed the "low-cost" Diogenes Project and studied primary habitats;

realized several conversions and participated in competitions (projects include Worb railway station, with engineers Patrick Gartmann and Jürg Conzett, New Apostolic Church in Zuchwil, and "Welle von Bern"/ Swiss Federal Railway flyover, 2001–2005). Page 212.

Marcel Meili, architect, was born in Küsnacht (Zurich) in 1953. Studied architecture at the Swiss Federal Institute of Technology Zurich, graduated in 1980. 1981–1983 assistant professor at the institute gra/ETH (Swiss Federal Institute of Technology Zurich), 1993–1995 guest professor. Founded the architectural firm Meili & Peter in Zurich in 1987, since then also lecturer in Switzerland and the US. Projects/buildings: esp. interiors, single-family house in Wallisellen, wooden bridge in Murau (Austria) 1993–1995, Park Hyatt Hotel 1993–97, ABS-Bank Olten 1990 etc. Page 198.

Christian Menn, civil engineer (specialist for bridge design), was born in Zillis (Grisons) in 1927. Studied at the Swiss Federal Institute of Technology Zurich, graduated in 1950. After several years of professional practice, returned to the Swiss Federal Institute of Technology Zurich in 1953, doctorate in 1956. Worked for Société Dumez in Paris on the construction of the UNESCO building. Founded his own engineering consultancy in Chur and Zurich in 1957. Professor of Structural Engineering at the Swiss Federal Institute of Technology Zurich 1971–1992. In addition, he authored expertises, brought forth designs, and served on juries for bridge design competitions. Consulting engineer in many countries since 1992. Projects include: arched bridges in Grisons, 1958–1968; Rhine bridge at Reichenau, 1963; Felsenau bridge in Bern, 1975; Ganter bridge (Simplonstrasse), 1983; Sunniberg bridge near Klosters, 1998; Bridge over the Charles River, Boston, 2002. Page 208.

Bruno Monguzzi, graphic artist and typographer, was born in Mendrisio (Ticino) in 1941. Studied graphic design in Geneva, photography and psychology of perception in London. Staff member of Studio Boggeri, Milano, from 1961 onwards. Worked on 9 pavilions of the Expo 67 in Montreal. Returned to Milan in 1968, designed books, posters, and exhibitions. Bodoni Award 1971, further high-level awards in the US, Europe, and Japan. Collaboration with the Museo cantonale d'arte Lugano for many years. Extensive teaching positions in Lugano and Mendrisio, guest lecturer at numerous US universities. Lives in Meride (Ticino). Page 136.

Olivier Morand, architect, was born in Geneva in 1965. Studied architecture at the École Polytechnique Fédérale de Lausanne, graduated in 1990. As a student, internships with Atelier Cube in Lausanne. After graduation, worked as an architect at the Fondation Braillard in Geneva and the firm Arène-Edeikins Architectes in Paris. Founded the firm bmv in Geneva in collaboration with Hani Buri and Nicolas Vaucher in 1993. Assistant professor at the École Polytechnique Fédérale de Lausanne, professor in Geneva. Page 202.

Jacob Müller, designer, inventor, architect, was born in Zurich in 1905 and died in Ronco (Ticino) in 1998. Apprenticeship as a carpenter in Zurich, attended the Kunstgewerbeschule (School of Arts and Crafts) in Stuttgart, followed by an extended nomadic period of travels. Founded his own carpenter's workshop in Zurich in 1933. Initiated and co-founded the "Werkgenossenschaft Wohnhilfe" (Work Cooperative Housing Support) in 1945 and the "Schweizerische Arbeitsgemeinschaft Gestaltendes Handwerk" (Swiss Working Group Design Crafts) in 1947. Developed a series of modular furniture and standardized furniture, for instance the folding "Plio" in 1948. Numerous inventions and patents in the field of wood technology. Regular publications in the wood trade press since the early 1950s, gradual relocation to Ronco (Ticino), and dedicated work as an architect. Page 26.

Josef Müller-Brockmann, graphic designer, typographer, writer, and teacher, was born in Rapperswil (St. Gallen) in 1914 and died in Zurich in 1996. Leading theoretician and practitioner of Swiss typography. Began working as a graphic artist in 1952, clients included companies such as Porzellanmanufaktur Rosenthal AG in Selb (Germany), IBM Europe, Olivetti, and the Schweizerischen Bundesbahnen (Swiss Federal Railways). Taught at the Kunstgewerbeschule (School of Arts and Crafts) Zurich from 1957 to 1960 and at the Hochschule für Gestaltung Ulm (Ulm School of Design) in 1963. Page 64.

Yves Netzhammer, artist, was born in Schaffhausen in 1970. Training as an architectural draftsman, then studied visual design at the School of Art and Design Zurich, graduated in 1995. Since then, numerous video animations and installations as well as thematic visual works for books and newspapers. Awards and scholarships for several projects, numerous group and solo exhibitions. Page 264.

Neue Werkstatt: see Christoph Dietlicher, Thomas Drack, Andreas Giupponi.

Hamper von Niederhäusern, inventor and instrument maker, was born in Burgdorf in 1946. Training as an artistic blacksmith, Vocational School Bern. Took courses at the Kunstgewerbeschule (School of Art and Design) Bern, first major projects included works for Expo 64, blacksmith works for historic preservation projects. Employee of the zoological gardens in Zurich. After a period in southwest France, he worked as an offset printer; developments in laboratory design. Has his own metalworking shop today in Kaiserstuhl (Aargau). Designs and produces furniture, metalworks for buildings, and musical instruments. After independent study, built the first tuned sounding beams for xylophones and marimbas in 1986. Has his own instrument-building studio VONIE. Development and construction of the first floor xylophone Xala of wood and metal, in collaboration with Ania Losinger, 1998/99. Lives in Kaiserstuhl. Page 222.

Michel Nigg, architect and designer, was born in Zurich in 1970. Trained as an architectural draftsman and attended vocational school in Zurich. He graduated from the Zurich School of Design with a degree in Space and Product Design in 1996. Co-founded and headed the architecture and design company "Zenaro GmbH" in 1997. He is the co-proprietor of the "Nigg Architektur GmbH" since 2003. Many works in the residential and the retail sector. Designed interiors of restaurants, show rooms, trade fair stands, and exhibitions. Various other design projects including individual furniture pieces for, among others, kitchens, bathrooms, entranceways, and reception areas, including the shelving unit "Plattenspiel", created in 1999 with Markus Huber. Page 216.

Norm: see Dimitri Bruni and Manuel Krebs.

Siegfried Odermatt, was born in Neuheim (Zug) in 1926. He was self-taught when he entered the profession; worked for an industrial graphic designer and for the painter and graphic artist Hans Falk. Has worked as an independent graphic designer since 1950. Designed posters, brochures, leaflets, packaging, and books for numerous clients, among others: architectural and pharmaceutical industries, electric appliance companies, and insurance agencies. Founded a joint studio with Rosmarie Tissi in 1986. Participated in exhibitions worldwide (Europe, USA, Japan), received several distinctions. Lives in Zurich. Page 212.

Markus Peter, architect, was born in Zurich in 1957. After an apprenticeship as a technical draftsperson he studied architecture at the Technikum Winterthur (now Zurich University of Applied Sciences), and graduated in 1984. Worked for Dolf Schnebli in Zurich from 1985 to 1986. From 1986 to 1988 he was an assistant to Mario Campi at the Swiss Federal Institute of Technology, Zurich. Since 1987 has his own office "Meili & Peter" with Marcel Meili in Zurich. From 1993–1995 visiting professor at the Swiss Federal Institute of Technology, Zurich. Projects and buildings: esp. interior design, single-family houses in Wallisellen, wooden bridges in Murau, (Austria) 1993–1995, Park Hyatt Hotel 1993-1997, ABS-Bank in Olten 1990, etc. Page 198.

Zuzana Ponicanova, textile designer, was born in Nitra, Slovakia in 1971. 1985-1989 Kunstgewerbeschule (School of Arts and Crafts), Bratislava, 1989–1992 University of Fine Arts, Bratislava (art courses), 1992–1995 Studied textile design at the Zurich School of Design and graduated in 1995. 1996-2000 graduate studies at the Zurich School of Design, wrote her thesis on "The Theory of Design and Art", graduated in 2000. Since 2000 has taught courses at the School of Art and Design Zurich, after 2000 collaboration with Patricia Collenberg under the Label Collenberg/ Ponicanova. Lives and works in Zurich. Page 226.

Tania Prill, graphic designer, was born in Hamburg in 1969. Studied visual communications at the University of Art, Bremen and at the School of Art and Design Zurich. 2005 Executive Master: Art | Design + Innovation, Basel School of Art and Design. Since 2001 works with Alberto Vieceli. Founded the Zurich Studio Prill & Vieceli with an emphasis on typography and book design. Since 1997 teaches at Universities in Lucerne, Bern, and Zurich, since 2004 visiting and interim professorships of graphic design at the State Academy of Design Karlsruhe. Lives and works in Zurich. Page 225.

Michel Rappaport, architect, was born in Zurich in 1962. Studied architecture at the Swiss Federal Institute of Technology Zurich, graduated in 1989. 1990–1991 worked for Pfister + Schiess Architects, Zurich, then at Stutz + Bolt Architects, Zurich 1991-1998. Since 1995 has a joint firm (B.E.R.G. Architects) with Christoph Elsener and Sibylle Bucher in Zurich. Page 252.

Alois Rasser, designer and interior designer, was born in Zug in 1948. Trained as a carpenter, afterwards studied Interior Design at the Kunstgewerbeschule (School of Arts and Crafts) Zurich under Hansruedi Vontobel, Alf Aebersold, and Willi Guhl 1968-1972. Opened own studio in 1974 in Baar and taught the works and design teachers seminar at St. Michael in Zug. In 1990 founded the studio and gallery for industrial art in Zug, DAS DA. Projects and design work (interior design commissions: restaurant, living room; furniture for interiors and exteriors, color conceptualizations). Lecturer at the design department of the Technical College for Technology and Design Zug, since 1997. Page 121.

Patrick Reymond, architect and designer, was born in Biel in 1962. Sudied architecture and design at the School Athenaeum Architecture & Design, Lausanne. During his studies did several internships at home and abroad. Graduated in 1990. Founded the Atelier Oï in La Neuveville with Armand Louis and Aurel Aebi in 1991. Many works in the fields of architecture, design, and scenography; clients include: Ikea, Wogg, Roethlisberger, Swatch, B & B Italia, Expo 02/La Neuveville, Rosenthal, etc. Teaches at the design department of the University of art and design Lausanne. Page 260.

Jean Robert, visual designer, was born in La Chaux-de-Fonds in 1945. Studied at the École des Arts et Métiers in La Chaux-de-Fonds. Spent several years working as a visual designer in Italy and England. In 1976 he founded the "Robert & Durrer" studio in Zurich with his partner Käti Durrer. 1983-1989 worked as a consultant and managed the entire conception and design of the Swatch collections. Since 1990 numerous commissions for various cultural institutions designing posters, invitation cards, catalogues, among others for the Kunsthalle Zurich, the Fotomuseum Winterthur, and the Fondation Beyeler. In addition designed several books. Has taught in Switzerland and abroad. Page 236.

Benedikt Rohner, was born in Pratteln in 1925, died in Zurich in 2000. Studied at the Kunstgewerbeschule (School of Arts and Crafts) Zurich under Wilhelm Kienzle and Willy Guhl, graduated in 1948. Opened an interior design company in Zurich, since 1960 continual collaboration with the cabinetmaker Ph. Oswald (including drawer cabinets). Has his own workshop for model making; designed exhibition displays including for IBM. Page 49.

Franz Romero, architect, was born in Zurich in 1951. Trained as an architectural draftsman in Zurich, followed by internships. Began his studies at the building engineering department of the Technikum Winterthur (now Zurich University of Applied Sciences), in 1974, graduated in 1977. Continued his studies at the Swiss Federal Institute of Technology Zurich, graduated in 1980. Study abroad at the Institute of Urban Architecture, New York. 1980–1981 worked as an independent architect, 1981–1985 worked in Theo Hotz's office in Zurich. Several projects and buildings: commercial house in Zurich, engine shed for the Swiss Postal Services, in Mülligen etc. Founded his own architecture firm in 1986. Markus Schaefle became his business partner in 1988. Assistantships and guest professorships in Switzerland and abroad. Page 142.

Flora Ruchat-Roncati, architect, was born in Mendrisio (Ticino) in 1937. Studied at the Swiss Federal Institute of Technology Zurich, graduated in 1961. From 1962 to 1971, collaborated with the architects Aurelio Galfetti and Ivo Trümpy. Founded her own architecture firm in Riva San Vitale in 1971, since 1975 also one in Rome. From 1985 to 2002, professor at the Swiss Federal Institute of Technology Zurich. Since 1988, partnership with Dolf Schnebli and Tobias Ammann in Zurich and Agno. Worked in conjunction with several offices including La Transjurane (with Renato Salvi), Mario Botta, or Luigi Snozzi. Several public and private buidlings including the public swimming pool in Bellinzona 1967–1970 (with A. Galfetti and I. Trümpy), school buildings, residential buildings, administrative buildings etc. Page 133.

Nelly Rudin, graphic designer and artist, was born in Basel in 1928. Trained as a visual designer at the Kunstgewerbeschule (School of Arts and Crafts) Basel. In 1950 worked freelance at chemical company J. R. Geigy's graphic studio in Basel, 1953–56 freelance work for various advertising companies in Zurich; 1957–1962 own studio in Zurich; in 1964 she began work as a freelance artist creating paintings and illustrations. Lives and works in Uitikon (Zurich). Page 92.

Marc Saugey, architect, was born in Collonge-Bellerive (Geneva) in 1908, died in Geneva in 1971. Studied architecture at the École des Beaux-Arts Geneva, graduated in 1926. Since 1931 membership in the Groupe pour l'architecture nouvelle à Genève. Worked vigorously in construction in Geneva, from 1967 onwards also in Spain and in Turkey. Founded the magazine "Architectures, formes, fonctions" with Alberto Sartoris and Anthony Krafft in 1956. From 1961 to 1970 he was a professor and director of the Urban Planning department, École d'architecture, at the University of Geneva. Page 82.

Colin Schaelli, designer, was born in Chur in 1980. Trained as an architectural draftsman under Peter Zumthor, followed by design studies at the School of Art and Design Zurich, graduated in 2006. Since 2004 he works as a Freelancer at FREITAG. Currently he also works for numerous businesses as a creative consultant.

Hans Ulrich Scherer, architect, was born in Klingnau (Aargau) in 1932, died in Gockhausen (Zurich) in 1966. Studied at the École des Beaux-Arts in Paris and at the Swiss Federal Institute of Technology Zurich. In 1958 he participated in the group exhibition "Brugg 2000". Known for his trademark terrace housing including those in Mühlehalde in Umiken (1963–1966). Page 268.

Adrian Schiess, artist, was born in Zurich in 1959. Several solo exhibitions in Switzerland and abroad. Participated in the 1990 Venice Biennale and the 1992 Documenta in Kassel. Lives in Southern France. Page 218

Smarch: see Beat Mathys, Ursula Stücheli.

Silvio Schmed, architect, was born in Chur in 1952, grew up in Trun (Grisons). Trained as an architectural draftsman under Andres Liesch in Zurich, followed by studies of interior design and product design at the Kunstgewerbeschule (School of Arts and Crafts) Zurich, graduated in 1978. From 1979 to 1987 worked as an architect of exhibitions at the Museum of Design Zurich. Designed furniture for the company Ph. Oswald AG. Since 1987 worked as an independent architect in Zurich, currently in collaboration with Arthur Rüegg. Page 166.

Ralph Schraivogel, graphic designer, was born in Lucerne in 1960. Studied graphic design at the Zurich School of Design (1978–1982), then worked as an independent graphic designer, predominantly for cultural institutions (including Filmpodium Zurich, Pro Helvetia, several museums). Has been a lecturer at the Zurich School of Design since 1992. Has received various international distinctions, numerous exhibitions. Lives and works in Zurich. Page 187.

Thomas Schregenberger, architect, was born in St. Gallen in 1950. Studied at the Städelschule of Fine Art Frankfurt and at the Architectural Association School of Architecture London. Taught at various domestic and foreign universities. Currently director of a drafting studio at the Institute for Architecture and Interior Design at the University of Liechtenstein in Vaduz. Several buildings including current designs for the restructuring of two industrial areas: the Hürlimann-Area in Zurich and the Turicum-Area in Uster. Lives and works in Zurich. Page 192.

Luigi Snozzi, Architect, was born in Mendrisio (Ticino) in 1932. Studied architecture at the Swiss Federal Institute of Technology Zurich, graduated in 1957. Employed by Peppo Brivio and Rino Tami. Since 1962 works freelance, until 1968 in a partnership with Livio Vacchini. Taught at the Swiss Federal Institute of Technology Zurich 1973–1975 and at the University of Geneva from 1980 to 1982. 1985–1997 he was a professor at the École Polytechnique Fédéral de Lausanne. Numerous teaching positions across Europe and in Latin America. Received the Wakker Award for the redesigning of the Monte Carasso in 1993. Lives and works in Locarno. Page 144.

Nadine Spengler, graphic designer, was born in Zurich in 1972. Studied at the School of Art and Design Zurich. Since 1995 works as an independent graphic designer and illustrator in Zurich. Co-founder and proprietor of Pipifax, a small store and publisher of artistic books and illustrated magazines. Works collaboratively with Franziska Burkhardt and Anna Albisetti, among others. Page 268.

Annette Spillmann, architect, was born in Zurich in 1969. Study of architecture at the Swiss Federal Institute of Technology Zurich, graduated in 2000. Founded partnership firm with Harald Echsle in Zurich in 2002. Page 266.

Georg Staehelin, graphic designer, was born in Basel in 1942. Studied graphic design under Armin Hofmann and Emil Ruder in Basel (1958–1963). Worked in several offices, including Gérard Ifert's office in Paris and Basel (1963–1964), at Total Design Amsterdam (1964–1966) and at Crosby/Fletcher/Forbes in London (1966–1969). 1969–1977, founder and director of Pentagram Design Zurich. Taught in the photography department at the Zurich School of Design (1977–1990). Founded his own studio for visual design in Ottenbach (Zurich) in 1977. Page 170.

Astrid Staufer, architect, was born in Lausanne in 1963. Study of architecture at the Swiss Federal Institute of Technology Zurich, graduated in 1989. Worked at the Meili & Peter firm 1990–1992, began working freelance in 1993, founded partnership firm with Thomas Hasler in 1994. Lecturer of constructive design at the Zurich University Winterthur since 1994. Offices in Zurich and Frauenfeld. Page 198.

Susanne Stauss, photographer, was born in Stuttgart in 1965. 1981 preparatory course at the Kunstgewerbeschule Zürich (School of Arts and Crafts Zurich), from 1982 onwards studied textiles in Zurich and graduated in 1987. 1987–1991 embroidery work in Germany and in Switzerland. 1991–1996 studied photography, and graduated from the Zurich School of Design. Specializes in artistic documentary photography. Lives in Zurich. Page 258.

Ursula Stücheli, architect, was born in Zurich in 1963. Study of architecture at the Swiss Federal Institute of Technology, graduated in 1988. Taught in Basel. In 1991 founded the architecture firm smarch with Beat Mathys. Developed the "low cost" Diogenes-Project and studied primary habitats, realization of several buildings, and participation in competition projects including: from 2001–2005 the Worb train station concourse with the engineers Patrick Gartmann and Jürg Conzett, the New Apostolical Church in Zuchwil, and the Swiss Federal Railway flyover in the western portal of the Bern central train station. Page 212.

Kurt Thut, Designer, was born in Möriken (Aargau) in 1931. Trained as a cabinetmaker, followed by the study of interior design at the Kunstgewerbeschule Zürich (School of Arts and Crafts Zurich) under Willy Guhl and Hans Fischli. Worked in Hans Fischli's firm. Opened his own architecture and interior design firm in 1961. Took over the family cabinetmaking business. Designed, manufactured, and developed furniture. Further furniture designs in light metal, including collapsible furniture for urban nomads such as the "Scherenbett" (2001) and the "Folienschrank". Co-founder of the designer trio "Swiss Design" along with Hans Eichenberger and Robert Haussmann. Since 1989 close collaboration with Heinz Ryffel for the production and distribution company Seledue. Page 164.

Niklaus Troxler, graphic designer, was born in Willisau in 1947. Trained as a typographer 1963–1967, study of graphic design at the Kunstgewerbeschule (School of Arts and Crafts) Lucerne 1967–1971. Artistic Director at Hollenstein Création in Paris 1971–1972. He founded his own graphic design studio in Willisau in 1973. Since 1975 he has run the Jazz festival in Willisau. He has been a professor at the State

Academy of Graphic Arts in Stuttgart since 1998. Page 146.

Ivo Trümpy, architect, was born in Lugano in 1937, grew up in Bellinzona. Graduated with a degree as an architect-technician in 1958. Staff member at the Pagnamenta firm 1959-1961, collaboration with Aurelio Galfetti and Flora Ruchat 1962-1970. Page 133.

Livio Vacchini, architect, was born in Lugano in 1933. Study of architecture at the Swiss Federal Institute of Technology Zurich 1953–1958, graduated under Rino Tami. Employed in Stockholm and Paris 1959-1960. Since 1961 works freelance in Locarno, 1962–68 partnership with Luigi Snozzi. A series of significant buildings in Ticino: apartments, offices, businesses, and school buildings. Lives and works in Locarno. Page 144.

Nicolas Vaucher, architect, was born in Geneva in 1965. Study of architecture at the École Polytechnique Fédéral de Lausanne and at the Swiss Federal Institute of Technology Zurich, graduated in 1991. During his studies, interned at Atelier Cube in Lausanne, Dunning & Versteegh in Geneva, and Peter Eisenmann in New York. 1991–1993, worked for François Roche in Paris and as a freelance architect in France and in Switzerland. In 1993 he co-founded the firm bmv in Geneva with Olivier Morand and Hani Buri. Assistant teacher at the École Polytechnique Fédéral de Lausanne 1995-1998. Editorial work for various specialized magazines. Page 202.

Regula Verdet, designer, born in Stäfa (Zurich) in 1952. Training as a handicraft teacher at the Zurich School of Design, lived abroad in Germany and France, academic travels to Lithuania and Hungary. She currently works as a textiles teacher, designer, and businesswoman. She lives and works in Guarda (Engadin). Page 206

Alberto Vieceli, graphic designer, was born in Zurich in 1965. Studied graphic design at the Zurich School of Design, graduated in 1992. Staff member at the Atelier Polly Bertram and Jul Keyser, Zurich (1988–1992) and at the Atelier Lars Müller, Baden (1994–1996). Opened his own studio in Zurich in 1996. Has worked collaboratively with Tania Prill since 2001. Founded the Zurich Studio Prill & Vieceli that specializes in typography and book design. Taught at the Lucerne School of Art and Design in 2004. Several publications. Page 256

Pascal Vincent, architect, was born in Geneva in 1964. Graduated with a degree in architecture from the École Polytechnique Fédéral de Lausanne in 1989. 1990–1996, staff member at Atelier 5 in Bern, 1996 founded the architecture firm Aebi & Vincent with Pascal Vincent in Bern. Public and private projects: housing developments, single family houses, various renovation projects, Valient Bank in Bern, renovation of the Federal Building (Bundeshaus), etc. Page 194

Carlo L. Vivarelli, was born in Zurich in 1919, died in Zurich 1989. Trained as a graphic designer and studied at the Kunstgewerbeschule Zürich (School of Arts and Crafts Zurich), 1939 Paris, 1946 Artistic Director of the Studio Boggeri, Milan. He had his own studio in Zurich since 1947. Co-editor of the "Die neue Grafik" ("New Graphic Design") (with Lohse, Neuburg, Müller-Brockmann). Worked as a painter and sculptor since 1950. Won numerous distinctions and competitions. Page 21.

Klaus Vogt, architect and designer, was born in Winterthur in 1938, raised in Zurich. After completing a naval construction training in Meilen, on lake Zurich from 1952 to 1956, he attended an interior design course at the Kunstgewerbeschule Zürich (School of Arts and Crafts Zurich, under Willy Guhl, Hans Fischli), 1962–1968 staff member at Dolf Schnebli architects firm, 1966–1970 assistant to Professor Bernhard Hoesli, in the architecture department of the Swiss Federal Institute of Technology Zurich. 1974–2003 lecturer of Analysis, Design and Construction in the architecture department of the Basel School of Art and Design. Visiting lecturer at the Swiss Federal Institute of Technology Zurich 1981–1983. Opened his own architecture firm in 1968, since 1974 in partnership with Benno Fosco and Jacqueline Fosco-Oppenheim. Lives in Scherz (Aargau). Page 124.

Daniel Volkart, visual designer, was born in Winterthur in 1959. Studied at the Kunstgewerbeschule Zürich (School of Arts and Crafts Zurich). Works as a visual designer in Zurich and teaches at the School of Art and Design Zurich and at the Lucerne school of Art and Design. Lives in Zurich. Page 196.

Hanspeter Weidmann, designer, was born in Basel in 1958. First studied law and later design at the Basel School of Art and Design. Founded his own studio for product, furniture, and interior design in Basel in 1986. Production and distribution of his own collection in Switzerland, licensed productions in Italy and Germany. Numerous national and international exhibitions. Lives and works in Basel. Page 154.

Heidi Wenger, architect, was born as Heidi Dellberg in Brig in 1926. 1946–1952 study of architecture at the Swiss Federal Institute of Technology Zurich, formed a partnership architecture firm in Brig with Peter Wenger in 1952. Taught in Nanjing, China in 1983. Expert consultant for Franz Füeg in 1986, École Polytechnique Fédéral de Lausanne. Alongside architecture she has numerous years of experience with writing and literature. Buildings with P. Wenger: among others the vacation home Trigon, the adult education center Tramelan 1987–1991. Lives in Brig. Page 50.

Peter Wenger, architect, was born in Basel in 1923. After a two-year study of machine technology he studied architecture at the Swiss Federal Institute of Technology Zurich, graduated in 1951. Partnership with Heidi Wenger in Brig since 1952. Their specialty: developing public structures as living space. Several teaching positions in Switzerland and abroad including the Swiss Federal Institute of Technology Zurich, in 1983 at the Southeast

University Nanjing in China and at the Slovak Technical University Bratislava. Works also with photography and sculpture (including tensegrity structures). Lives in Brig. Page 50.

Trix Wetter, graphic designer, was born in Zurich in 1947. Preparatory course at the Kunstgewerbeschule (School of Arts and Crafts) Bern from 1963 to 1964, studied graphic design at the Kunstgewerbeschule Biel 1964–1968. 1968–1974 she worked for the Hablützel & Jaquet studio in Bern, with a stint as a packaging designer for the advertising agency SNIP in Paris. Freelances since 1975, with a focus on art and architecture including magazine layout, posters, catalogues, and architectural books. Page 236.

Hannes Wettstein, designer, was born in Ascona (Ticino) in 1958. 1982-1988 he freelanced for the Büro für Gestaltung in Zurich. Founded the Edition "Diesseits" (mini-series and unique pieces) in 1989; 1989–1991, co-proprietor and head of the 3-D division of the design agency Eclat Erlenbach; 1991 founded zed and zednet-work in Zurich, bringing together artists from all design disciplines (areas of activity: public and private spaces, lighting, furniture, object design, strategy and research). Several teaching posts: 1991–1996, lecturer at the Swiss Federal Institute of Technology Zurich; 1990–1995, visiting lecturer at several Swiss and foreign academies and schools; 1994, professor at the State Academy of Design Karlsruhe; since 2002, lecturer at the School of Art and Design Zurich. Lives and works in Zurich. Page 181.

Gerold Wiederin, graduate engineer, architect, has a firm in Vienna. Worked on the design and realization of the Forum 3 building on the Novartis campus in Basel. Page 240.

Martin Woodtli graphic designer, was born in Bern in 1971. Studied first at the School of Visual Arts Bern and Biel (Bern) and subsequently visual communication at the School of Art and Design Zurich, graduated in 1998. Lived in New York from 1998–1999, returned to Zurich and founded his own graphic design studio. Works in both free and applied design. Since 2001, lecturer at the Lucerne School of Art and Design and at several institutions in Switzerland and abroad. Page 272.

Ruedi Wyss, was born in Zofingen in 1949, studied graphic design in Biel, initiator and organizer of contemporary music (including Festival Taktlos, the concert series Tonart, both in Bern). Currently Department Head and lecturer at the School of Art and Design Zurich. Various international exhibitions and distinctions. Lives in Zurich. Page 167.

Xala: see Ania Losinger and Hamper von Niederhäusern.

SELECTED BIBLIOGRAPHY

Architecture

(Aebi und Vincent): Aebi und Vincent Architects Colorfulness Publishing (Hg.): *Native Modernism, Switzerland.* Beijing, China 2004

Bill, Max: *Wiederaufbau.* Erlenbach-Zurich 1945 (Verlag für Architektur)

Bill, Max: *Robert Maillart.* Zurich 1949/1955/1965 (Girsberger)

Bill, Max: *Form. Eine Bilanz des Formschaffens um die Mitte des 20. Jahrhunderts.* Basel 1952 (Wenk)

(Bill) Frei, Hans: *Konkrete Architektur? Über Max Bill als Architekt.* Baden 1991 (Lars Müller Publishers)

(Bill) Niggli-Verlag (Hg.): Max Bill: *Typografie – Reklame – Buchgestaltung.* Sulgen 1999 (Niggli)

Billington, David P.: *The Art of Structural Design. A Swiss Legacy.* Princeton 2003 (Princeton University Art Museum)

Botta, Mario, Zardini, Mirko: *Aurelio Galfetti, Einführung.* Berlin 1989

Burckhardt, Lucius/Beutler, Urs: Terrassenhäuser. Winterthur 1969 (Verlag Werk)

Bürkle, J. Christoph: *Gigon/Guyer Architekten. Arbeiten 1989–2000.* Sulgen 2000 (Niggli)

Cabalzar, A./Caminada, G.A. u. a.: *Gion A. Caminada. Stiva da Morts – vom Nutzen der Architektur.* Zurich 2003 (gta)

Cavadini, Luigi: *Castelgrande a Bellinzona.* Lugano 1993 (Casagrande-Fidia-Sapiens)

Conzett, Jürg/Mostafavi, Moshen/Reichlin, Bruno: *Structure as Space.* London 2006 (AA)

(Clemens) Ehrat, Fredi, Helfenstein, Heinrich (Hg.): *Das Kino Studio 4. Eine Dokumentation über eine Raumgestaltung von Roman Clemens.* Zurich 1994 (self published)

Dechau, Wilfried: *Traversiner Steg.* Berlin 2006 (Wasmuth)

Gigon, A./Guyer, M./Wirz, H.(ed.): *Projekte Gigon/Guyer.* Lucerne 2004 (Quart)

Gilbert, Mark et al. (ed.): *Construction Intention Detail. Five Projects from Five Swiss Architects (Meili & Peter, Burkhalter & Sumi, Diener & Diener, Herzog & de Meuron, Peter Zumthor).* Zurich 1994 (Artemis)

Gimmi, Karin et al. (ed.): *Max Bill Architect.* Barcelona 2004, (GG 29/30)

(Gisel) Maurer, Bruno/Oechslin, Werner (ed.): *Ernst Gisel Architekt.* Zurich 1993 (gta)

(Gisel) Scheidegger, Ernst (ed.): *Ernst Gisel Architekt, Aquarelle und Zeichnungen,* Zurich 1990 (Scheidegger & Spiess)

Haller, Fritz: *Fritz Haller. Bauen und Forschen.* Solothurn 1988 (Kunstverein Solothurn)

(Hilfiker) *Hans Hilfiker, Ingenieur und Gestalter.* Catalogue, Museum für Gestaltung Zürich/Kunstgewerbemuseum (Reihe Schweizer Design Pioniere 1), Zurich 1984

Hossdorf, Heinz: *Das Erlebnis, Ingenieur zu sein.* Basel 2003 (Birkhäuser)

(Isler) Ramm, Eckehard/Schunck, Eberhard: *Heinz Isler – Schalen.* Stuttgart 1989 (Krämer)

Jehle Schulte-Strathaus, Ulrike (ed.): *Novartis Campus Forum 3. Diener, Federle, Wiederin.* Basel 2005 (Christoph Merian)

Jodidio, Philip: *CH Architecture in Switzerland.* Cologne 2006 (Taschen)

(Kerez) *Les échelles de la réalité. L'architecture de Christian Kerez.* Lausanne 2006 (Kat. EPFL)

Lichtenstein, Claude/Schwarz, Marc: Jacques Schader – Freudenberg. *A Masterpiece of European Architecture.* Baden 2003 (Lars Müller Publishers)

Moos von, Stanislaus/Bachmann, Jul: *New Directions in Swiss Architecture.* New York 1969 (Braziller)

Ruchat-Roncati Flora: *Exhibition Catalogue.* Zurich 1998 (Edition gta/ETHZ)

(Saugey) Saugey, Marc: Faces Nr. 21, *Journal d'architecture.* Carouge/Geneva, Fall 1991

(Saugey) Valérie Opériol et al. (ed.): *Le Cinéma Manhattan à Genève. La révélation d'un espace.* Geneva 1992 (NLDA)

Schlorhaufer, Bettina (ed.): *Cul zuffel a l'aura dado. Gion A. Caminada.* Lucerne 2005 (Quart)

Snozzi, Luigi: *Monte Carasso. Die Wiederfindung des Ortes. La reinvenzione del sito.* Basel/Boston/Berlin 1995 (Birkhäuser)

(Snozzi) Lichtenstein, Claude: *Luigi Snozzi.* Basel/Boston/Berlin 1997 (Birkhäuser)

Steinmann, Martin (ed.): *Tendenzen der neueren Tessiner Architektur.* Zurich 1975 (Katalog gta, ETHZ)

Werner, Frank (ed.): *Aurelio Galfetti: Castelgrande in Bellinzona.* Berlin 1992 (Ernst & Sohn)

Zschokke, Walter/Hanak, Michael (ed.): *Nachkriegsmoderne Schweiz. Werner Frey, Franz Füeg, Jacques Schader, Jakob Zweifel.* Basel/Boston/Berlin 2001 (Birkhäuser)

Graphic Design

Bignens, Christoph: *Swiss Style: Die grosse Zeit der Gebrauchsgrafik in der Schweiz 1914–1964.* Zurich 2000 (Chronos)

Bruggisser, Thomas/Fries, Michel (ed.): *Benzin. Junge Schweizer Grafik.* Baden 2000 (Lars Müller Publishers)

Bruggisser, Thomas/Michel Fries (ed.): *Super. Welcome to Graphic Wonderland.* Berlin 2003 (Die Gestalten)

Ernst, Meret/Eggenberger, Christian (ed.): *Design Suisse.* Zurich 2006 (Hochparterre, Scheidegger & Spiess)

Frutiger, Adrian: *Buch der Schriften. Anleitungen für Schriftenentwerfer.* Wiesbaden 2005 (Marix)

Gerstner, Karl: *Programme gestalten.* Teufen 1963 (Niggli)

Hochuli, Jost: *Buchgestaltung in der Schweiz.* Zurich 1993 (Pro Helvetia)

Hofmann, Armin: *Methodik der Form- und Bildgestaltung. Aufbau, Synthese, Anwendung.* Sulgen/Zurich 1965, 5th ed. 2005 (Niggli)

(Hofmann) Wichmann, Hans (ed.): *Armin Hofmann – his work, quest and philosophy.* Basel 1989 (Birkhäuser)

Hollis, Richard: *Swiss graphic design: the origins and growth of an international style, 1920–1965. London 2006 (Laurence King)*

Kenner, Markus/Lichtenstein, Claude (ed.). *Follow the Signs. Party-Flyers in Zürich.* Catalog, Edition Museum für Gestaltung Zürich. Zurich 2000

Klanten, R./Mischler, M (ed.): *Woodtli.* Berlin 2001 (Die Gestalten)

Klanten, Robert (ed.): *Büro Destruct.* Berlin 1999 (Die Gestalten)

Klanten, Robert (ed.): *Büro Destruct II.* Berlin 2003 (Die Gestalten)

Lohse, Richard Paul: *Neue Ausstellungsgestaltung/ Nouvelles conceptions de l'exposition/New Design in Exhibitions.* Zurich 1953 (Verlag für Architektur)

Lutz, Hans-Rudolf: *Edmonton Journal.* Zurich 1977

Lutz, Hans-Rudolf: *Ausbildung in typografischer Gestaltung.* Zurich 1987/1989/1996) (Verlag Hans-Rudolf Lutz)

Lutz, Hans-Rudolf: *Typoundso.* Zurich 1987/1996 (Verlag Hans-Rudolf Lutz)

Monguzzi, Bruno: *Lo studio Boggeri 1933/1981.* Milano 1981 (Electa)

(Monguzzi) Nunoo-Quarcoo, Franc: *Bruno Monguzzi. A Designer's Perspective.* Baltimore 1998 (Kat. Fine Arts Gallery, University of Maryland)

Müller, Lars: *Helvetica. Homage to a Typeface.* Baden 2002 (Lars Müller Publishers)

Müller, Lars: *Josef Müller-Brockmann Gestalter.* Baden 1994 (Lars Müller Publishers)

Müller-Brockmann, Josef: *The Graphic Artist and his Design Problems. Teufen 1961 (Niggli)*

Norm: *The Things.* Zurich 2002 (Norm/Die Gestalten, Berlin)

Odermatt, Siegfried (ed.): *100 + 3 Schweizer Plakate, ausgewählt von Siegfried Odermatt.* Zurich 1998 (Waser)

Troxler, Niklaus: *Jazz Blvd.*. Baden 1999 (Lars Müller Publishers)

Wyss, Ruedi, Kraut, Peter (ed.): *13/11/1980: Archie Shepp/ Jasata – 20/12/2006: Pan Sonic/Alter Ego. Eine Materialsammlung zu den 580 Konzerten von taktlos bern und tonart bern.* Bern 2006 (edition taktlos bern)

Poster Collection, Edition Museum für Gestaltung Zürich, Plakatsammlung/Lars Müller Publishers: 07: *Armin Hofmann* (2003), 09: *Ralph Schraivogel* (2003), 11: *Handmade* (2005)

Design

Crivelli, Patrizia/Münch, Andreas (ed.): *Made in Switzerland.* Exhibition Catalog. Bern 1997 (Bundesamt für Kultur)

Erni, Peter: *Die gute Form. Eine Aktion des Schweizerischen Werkbundes.* Baden 1983 (LIT, Lars Müller)

Ernst, Meret et al.: *Designsuisse.* Zurich 2006 (Hochparterre, Scheidegger & Spiess)

Frank, Beat/Lehmann, Andreas: Atelier Vorsprung – Beat Frank + Andreas Lehmann: 32 Möbelobjekte 1986–1988. Bern 1988 (Salchli)

(Guhl) *Willy Guhl – Gestalter und Lehrer.* Catalogue, Museum für Gestaltung Zürich (Reihe Schweizer Design-Pioniere 2), Zurich 1985

Heller, Martin/Windlin, Cornel (ed.): Universal. Catalogue, Edition Museum für Gestaltung Zürich), Zurich 1996

(Kienzle) *Wilhelm Kienzle.* Ausstellungskatalog Museum für Gestaltung Zürich, Reihe Schweizer Designpioniere 6. Zurich 1991

Lueg, Gabriele/Gantenbein, Köbi (ed.): *Swiss Made, Aktuelles Design aus der Schweiz, Ausstellungskatalog Museum für angewandte Kunst Köln.* Zurich 2001 (Hochparterre)

Meier, Andreas, Tschopp, Walter (ed.): *Ueli Berger.* Bern 1992 (Benteli)

Rüegg, Arthur (ed.): *Swiss Furniture and Interiors in the 20th Century.* Basel/Boston/Berlin 2002 (Birkhäuser)

Schilder Bär, Lotte/Wild, Norbert (Hg.): *Designland Schweiz. Gebrauchsgüterkultur im 20. Jahrhundert.* Zurich 2001 (Pro Helvetia)

Schwarz, Dieter: *Andreas Christen Werke 1958–1993.* Winterthur 1994 (Kunstmuseum)

(Thut)*Thut Möbel – 1953 bis heute.* Zurich 2001 (Verlag Hochparterre)

PHOTO CREDITS

Georg Aerni: 262pp.
Baltensweiler: 56pp.
Binia Bill: 38pp. (Max, Binia and Jakob Bill Foundation)
P. Bötschi: 117 (Metron AG)
Pino Brioschi: 133–135 (Archivio del Moderne, Mendrisio)
Susann Brütsch: 254
Centre d'iconographie genevoise: 82-86
Documentation photographique de la ville de Genève: 203
F. Engesser: 42, 44 right, 45, 45 (SBB Historic, Bern)
ETH Zurich, Chair for Architecture and
Digital Production: 275–277
Alberto Flammer: 144pp.
René Furer: 119
Yvonne Griss: 152pp. (archive Neue Werkstatt)
Max Hellstern: 98pp. (ETH Zurich/gta)
Heinrich Helfenstein: 22–25,198, 219–221
HGK Zurich, Design Collection (photography
Franz Xaver Jaggy): 31 left, 48, 120pp., 155
HGK Zurich, Poster Collection: 26, 41, 77, 79–81, 89, 91, 94,
138pp., 171pp., 188-190
Heinz Isler: 104–109
Samuel Jäggi: 214pp.
Thomas Jantscher: 246pp., 280-284
Valentin Jeck: 232pp.
Lili Kehl: 192pp.
Lehni AG: 114pp.
Kathrin Leuenberger: 226
Claude Lichtenstein: 35 top, 116 right, 118, 182-185, 205,
208, 210 bottom right, 211, 231, 233
Beat Mathys: 213
Christian Menn: 210 top left, 210 top right
Martin Möll: 222pp.
Michel Nigg: 216
Erica Overmeer: 242
Doris Quarella: 115
Urs Ramseier: 163
Christian Richters: 240
Jakob Schildknecht: 32, 34, 35 bottom, 36
(archive H: Danzeisen)
Marc Schwarz: 96, 206
Urs Siegenthaler: 101 (ETH Zurich/gta)
Margherita Spiluttini: 199
Alexander Troehler: 49, 166 (Ph. Oswald)
Thut & Knup: 164pp.
Klaus Vogt: 124
Peter Wenger: 50–55
Vaterlaus: 116 left
Michael Wolgensinger: 62, 97
Reinhard Zimmermann: 252pp.
Archive Josef Müller-Brockmann: 68
Archive Franz Romero: 58–60, 142pp.
Archive Magda Vogel: 158pp.

Drawing on p. 230: Gregor Katz, Basel.

Photographs not mentioned here
are from the archives of the respective
designers

Acknowledgments
I owe a debt of gratitude to the Kornhausforum Bern under the direction of Claudia Rosiny for its willingness to open these questions for discussion in the context of the exhibition of the same name. I thank my fellow team members, Sabine von Fischer and Marc Schwarz, for their close cooperation and for countless impulses during our collaboration on this topic. My thanks go to all the designers and/or their representatives for the enriching encounters, the conversations that provided inspiration in many ways, and for all the assistance they have afforded. My appreciation also includes René Furer, Annemarie Hürlimann, Daniel Stettler, Anne Vonèche, and Conradin Wolf. Marianne Unternährer Pickard and her colleagues Frank Mayer, Jakob Steib, and Max Steiger of the Zurich University of Applied Sciences Winterthur (Architecture Department) provided valuable support with the building of the architectural models. I am very thankful to Atelier Integral Lars Müller and Publishers for, once again, very intense and enriching collaboration. Claude Lichtenstein

On the author
Claude Lichtenstein, born 1949 in Zurich. Formation as an architect (ETH Zurich), professional activities as a curator, teacher, and author. Collaboration with Lars Müller on these publications: *Streamlined* (1993, with Franz Engler), *Bruno Munari* (1995, with Alfredo W. Häberli), *R. Buckminster Fuller* (2 vol. 1999/2001, with Joachim Krausse), *As Found/Independent Group* (2001, with Thomas Schregenberger), *Freudenberg-Schoolhouse* (DVD and book 2003, with Marc Schwarz). Lives in Zurich.

Design: Integral Lars Müller/
Séverine Mailler, Lars Müller
Translation: Sandra Lustig, Michael Dills, Cass Hirsh, Laura Radosh, Josee Schick (all Berlin)
Collaboration on the artists' biographies: Andrea Baur
Lithography: Ast & Jakob, Vetsch AG, Köniz
Printing: Jütte-Messedruck GmbH, Leipzig
Binding: Kunst- und Verlagsbuchbinderei GmbH, Leipzig

Published on the occasion of the exhibition *Playfully Rigid* in the Kornhausforum Bern

Lars Müller Publishers
CH–5400 Baden/Switzerland
books@lars-muller.ch
www.lars-muller-publishers.com

English edition
ISBN 13: 978-3-03778-090-9
ISBN 10: 3-03778-090-8

German edition
ISBN 13: 978-3-03778-089-3
ISBN 10: 3-03778-089-4

Printed in Germany

With the support of

swiss arts council
prohelvetia